Stochastic processes for water scientists

Stochastic processes for water scientists

Developments and Applications

Robin T. Clarke
Universidade Federal do Rio Grande Do Sul, Brazil

JOHN WILEY & SONS
Chichester · New York · Weinheim · Brisbane · Singapore · Toronto

National 01243 779777
International (+44) 1243 779777

e-mail (for orders and customer service enquiries): cs-books@wiley.co.uk
Visit our Home Page on http://www.wiley.co.uk
or http://www.wiley.com

Other Wiley Editorial Offices

John Wiley & Sons, Inc., 605 Third Avenue,
New York, NY 10158-0012, USA

Wiley-VCH Verlag GmbH, Pappelallee 3,
D-69469 Weinheim, Germany

Jacaranda Wiley Ltd, 33 Park Road, Milton,
Queensland 4064, Australia

John Wiley & Sons (Asia) Pte Ltd, 2 Clementi Loop #02-01,
Jin Xing Distripark, Singapore 129809

John Wiley & Sons (Canada) Ltd, 22 Worcester Road,
Rexdale, Ontario M9W 1L1, Canada

Library of Congress Cataloging-in-Publication Data

Clarke, Robin T., 1934–
Stochastic processes for water scientists: developments and applications / Robin T. Clarke.
p. cm.
Includes bibliographical references and index.
ISBN 0-471-97348-3
1. Hydrology—Statistical methods. 2. Stochastic processes. I. Title.
GB656.2.S7C57 1998
551.48'07'2—dc21 97-26708
CIP

British Library Cataloguing in Publication Data

A catalogue record for this book is available from the British Library

ISBN 0-471-97348-3

Typeset in 10/12pt Times from the author's disks by Vision Typesetting, Manchester
Printed and bound by Antony Rowe Ltd, Eastbourne

This book is printed on acid-free paper responsibly manufactured from sustainable forestry, in which at least two trees are planted for each one used for paper production.

For my children

Harriet, Richard, Jessica, Thomas and Julia

Contents

Preface		ix
Chapter One	Models: deterministic, time series and stochastic	1
Chapter Two:	Mathematical preliminaries	11
Chapter Three	Markov chains: applications to modelling of daily rainfall	41
Chapter Four	More on Markov chains: storage	77
Chapter Five	Stochastic reservoir theory applied to rainfall–runoff modelling	95
Chapter Six	The Poisson and related processes	111
Chapter Seven	Random walks and diffusion processes	151
Chapter Eight	Some computational methods	163
References		175
Index		181

Preface

WHAT IS MEANT BY "WATER SCIENCES" IN THE TITLE?

This book describes applications of stochastic processes in some of those fields of applied physical science that are closely linked to the land phase of the hydrological cycle. Any book aspiring to the presentation of technical material is inevitably biased by the filter of the author's experience, and this book is no exception. Many of the applications described have a basis in the hydrological sciences, or in associated environmental studies where water transport plays an important role. Nor will the reader find any formal definition of stochastic processes in this book; for its purpose, it is sufficient to say that they describe the behaviour of variable quantities, capable of observation and measurement, having numerical values that cannot be predicted with certainty, and that are subject to probabilistic laws.

A HISTORICAL VIEW

As in other areas of applied science, the use of quantitative methods in the water sciences is a field in which a number of often distinct but sometimes overlapping areas can be identified. These sub-fields do not fall naturally into any historical sequence, but rather their development has been associated with the activities of outstanding workers in the field. Perhaps the most important of the sub-fields consists of those methods which may be described as deterministic; the deterministic approach, which was developed by St Venant, Darcy and others in the nineteenth century, and by Dooge, Philip and others in the twentieth, has flourished where hydraulic engineers, physicists and mathematicians studied the physical processes of channel flow and water movement in porous media, largely under laboratory or otherwise experimental conditions. Corresponding developments in limnology were Mortimer's studies of circulation and thermal regime of lakes. The deterministic models developed by these pioneers describe physical processes in terms of differential equations which, when solved subject to given initial and boundary conditions, yield predictions of future behaviour. Observations are compared with predictions, but differences between observation and prediction are not regarded as possessing any inherent probabilistic structure, nor are the uncertainties of definition of boundary conditions formulated in probabilistic terms. Deterministic methods, in which processes are described in terms of differential equations with coefficients having explicit physical interpretation, have particular importance where responses to changing environment are to be predicted, and their practical use in weather forecasting and other fields has been greatly stimulated in the second half of the twentieth century by the rapid growth of computer technology. Because of the nature of deterministic methods, however, uncertainties in predictions derived from deterministic models cannot be expressed in probabilistic terms (although data resampling methods will probably be increasingly used for this purpose as computer technology develops still further).

Where the response of river basins to rainfall was to be predicted, recognition of difficulties arising from uncertainties in boundary conditions led to modification of the deterministic approach in the form of linear systems theory and the calculation of unit hydrographs, Nash being pre-eminent in this field. It was reasoned that when a river basin is saturated after continued heavy rainfall, basin response could be treated as linear (in the sense that the response from the sum of two inputs is equal to the sum of their respective outputs). The success of the linear systems approach, and its modifications, is demonstrated by the fact that it remains widely used for the estimation of "design" floods. Uncertainty in the description of basin response remained a difficulty, however, since different rainfall events resulted in different unit hydrographs and, in the absence of any probabilistic component, tests for their mutual consistency were not straightforward. Furthermore, where rainfall events resulting in fully saturated conditions could not be identified, it was necessary to use empirical methods to allow for the "losses" that occurred by infiltration into the soil.

Recognition of the empiricisms in the separation of "losses", coupled with the growing development of computing power, stimulated the development of "conceptual models" of river basin behaviour, which represent it in terms of interlinked reservoirs modelling storage in surface depressions, soil and aquifer. Such models, associated originally with the names of O'Donnell and Dawdy but since extended by many people, have received much attention from hydrologists over the past 30 years. They were originally deterministic in character, although subsequently it was recognised that errors, caused by the spatial sampling of precipitation using rain gauges, introduced uncertainties which could best be described in probabilistic terms. Thus some workers came to regard the sequence of residuals (differences between observed and predicted runoff) as appropriate for time series analysis by the methods mentioned below, thereby regarding the residual sequences as random variables with probability distributions. In consequence, the constants ("parameters") of the rainfall–runoff models came to be regarded as unknown numerical quantities which had to be estimated from records of rainfall and runoff, and it was recognised that these estimates, being functions of the data used to estimate them, were also random variables. Similarly for predictions given by the "fitted" model. Over the past decade, techniques of rainfall–runoff modelling have assimilated still more of statistical methodology, including the concepts of likelihood and prior probability distributions for model parameters. Thus models that were originally deterministic in character have been metamorphosed into statistical constructs.

The rapid growth in computer technology enabled the more complex processes of basin-scale sediment removal, transport and deposition to be modelled also, using methods broadly similar to those of the rainfall–runoff models – except that it was necessary to include descriptions of the rates at which sediment became available for removal, and the rates at which it was removed by rain-splash and surface runoff. Even further extensions have been reported in which model descriptions of rainfall–runoff processes have been broadened to include the solution and transport of conservative and non-conservative chemical constituents.

The above paragraphs advance the view that models originally deterministic in character have been (and are being) transformed by the adoption of statistical methodology to describe uncertainties. Parallel with this transformation has been the growth in application of time series analysis methods for modelling hydrological time series. The statistical analysis of river flow sequences dates back at least to the early part of the twentieth century, when engineers such as Hazen, Hurst and others came to study records of river flow for the purpose of calculating storage volumes. Their analyses achieved remarkable penetration using very simple methods, and some decades passed before theoretical results by probabilists such as Lloyd caught up with results obtained much earlier by engineering intuition. The same engineers, working in the first quarter of the present century, realised the importance of Monte Carlo procedures and used them, long before powerful desk-top computers appeared. Thus the concepts of probability theory were exploited in this second strand of development in the water sciences, from its earliest days.

Probability theory was also the basis of two later developments. The first was the advance of the theory of storage, principally associated with the names of Moran and Lloyd. Subsequently, applications in the water sciences of time series analysis were given additional momentum by the classic text of Box and Jenkins, whose methods were used not only for predicting river flow, but also for the simulation of flow sequences for purposes of reservoir design, determination by simulation of reservoir operation rules, prediction of sediment yield and prediction of the frequency of extreme events. To a certain extent also, the explosive growth in application of time series methods – which still continues – eclipsed the theory of storage, since the latter, of necessity, greatly simplified the behaviour of water in storage, whilst the former was able to simulate more complex behaviour, although it had not the compactness and generality of mathematical solutions to storage problems.

Time series analysis of the kind developed by Box and Jenkins, and those who followed, has also been applied to the modelling of biological populations in fresh waters and the interdependence of predator numbers and prey. However, the complexities both of freshwater food chains themselves and of their interactions with physical and chemical environments, are such that time series analysis has been less fruitful than in the non-biological freshwater sciences. Time series analysis essentially provides tools for forecasting and for simulation without providing much enlightenment of the processes that give rise to sequences of observations; and it is enlightenment concerning their complex ecosystems that freshwater biologists seek.

Over the past 30–40 years, two of the detectable trends in the use of quantitative methods in the freshwater sciences have therefore been (a) a gradual synthesis of deterministic and statistical approaches to the description of river basin behaviour, as described above, and (b) the growth and wider application of time series analysis applied to sequences of hydrological, sedimentological and hydrochemical data. More recently, two further influences can be observed, both of which owe a good deal to the development of computer technology and the consequent ease with which numerical experimentation has become possible.

First, following the work of Robert May and others, very simple deterministic models have been shown to have great richness in generating sequences having the apparently random, "up-and-down" behaviour characteristic of fluctuations in the densities of biological populations. This has led to speculation concerning the true nature of "randomness" in such contexts, and concerning how far fluctuations in population densities can be explained in deterministic rather than in probabilistic terms. To this writer, such speculation loses sight of the fact that population densities under field conditions are almost always estimated by some form of sampling, yielding estimates subject to random errors with probabilistic structure.

The second influence has been the application of the theory of spatial processes and random field theory to the modelling of rainfall fields and of spatial variability in soil moisture and aquifer storage. Techniques used originally by mining engineers have been developed and put to extensive use by hydrologists, soil physicists and resource planners. A further impulse to their adoption has been the ever-increasing availability of remote-sensed data and satellite images.

WHAT THIS BOOK IS ABOUT

As argued above, time series analysis played an important role from the early days of water resource development, and this role has been extended with the growth of other sciences closely linked to the land phase of the water cycle. However, there are certain problems that are less conveniently dealt with by conventional time series methods. One such problem is where events occur at irregular time intervals; the greater part of the theory of time series requires that observations occur at points equally spaced in time, so that (for example) the irregular occurrences of rainfall events – which, in the final analysis, drive the land phase of the hydrological cycle – require some alternative treatment. The same problem can be extended to space of one or more dimensions, such as the irregular spatial occurrence of rain-cells observed in radar images of storm growth. A second problem for which time series analysis is less easily used is where one or more random variables describing the state of a phenomenon in nature fluctuate between physically imposed limits: examples are soil water content which commonly fluctuates between some physical lower limit, and an upper limit of saturation; volume of water stored in a unit area of forest canopy; mass of water vapour in a unit volume of air at given temperature and pressure.

Whilst it is possible to modify the commonly used methods of time series analysis for such problems, it is often more convenient to approach them in a different way. One approach to such problems is to describe the phenomenon in terms of random variables that assume permitted states with certain probabilities; the "state space" may be either discrete (days are either dry or wet; the number of floods occurring in any period must be some non-negative whole number) or continuous (the depth of water stored in a column of soil may take any value between an effective zero and saturation). Development of the phenomenon in time can then be conveniently described in terms of probabilities of transitions between the possible states. It is this approach that is propounded in the following pages.

Porto Alegre, RS, Brazil
June 1997

Chapter One

Models: deterministic, time-series and stochastic

1.1 A STORY

There was once a boy who enjoyed doing calculations, and who later would become a university professor of engineering. He asked his father, who happened to be one of the greatest scientists of the twentieth century, to set him a problem. This is how the boy described, much later, what happened (Einstein, 1971):

> More than half a century ago, when I was about 10 years old, I found great satisfaction in solving applied mathematical problems of the kind fitting a 10 year old boy. One day I pestered my father again with the request for a "calculation problem" as I may translate the German expression *Rechenaufgabe*. He just ended a period of work on Braunian movement, determination of the size of molecules and the beginnings of quantum theory, all those things which had absolutely no meaning for me then. So he took a long look at me as if he was not quite sure if he should dare to do what went through his head and finally decided to give it a try. He asked me, "How long will it take till the ground is wet if it rains at the rate of 10 mm per hour? Just take your time in finding out." It was the last sentence that got me worried – it meant that the problem was probably much more involved than just a simple division or some other simple operation. Besides, no matter how I tried to get started always there seemed to to be some of the most important information missing.
>
> After doodling for a while but getting nowhere, I admitted defeat and asked for help. This was naturally just what my father had been waiting for. He asked me first : "How does rain fall?" "In drops," I answered. "That is very important as you will see. How big is such a drop is really not important that we know exactly." I ventured "two millimeters." "Good. and how big is the splash from such a drop?" I guessed again: "a 10 millimeter circle. Now I know how to go about it." "Good," he said and puffed on his cigar seemingly agreeing with me. I went through the process of finding out how many 10 mm circles made a square meter and how long it would take for these drops to fall at the given rain [*sic*] and proudly showed him the result, just to be told: "That is all very nice, but it is not what I asked you. I asked you how long it takes to wet the area, not how

long it takes a given amount of rain to fall. You see, if you take the instant when half the area is wet, where will the next drop fall: Are you sure that it will hit a dry spot and wet another 1-cm circle?" I had to admit I did not know. That was the moment when I learned what the significance of probability is and that it takes actually many times longer to wet the area than I had calculated, because many drops are actually ineffective by hitting spots which were already wet. And from the fact that I still remember the entire episode proves that my father's experiment was successful in teaching me at the age of 10 the concept of probability.

The randomness of rain, and the need for methods based in probability theory where solutions are required to practical problems, is a theme underlying much of this book. Indeed the work by Calder (1986, 1996a, b) concerns a problem – the wetting of leaf surfaces by intercepted rainfall – is not dissimilar in character from the problem that the elder Einstein set his son. When, instead of rain falling at a constant rate of 10 mm per hour, variation in the hourly rate of rainfall is also taken into account, description of sequences of measured hourly rainfall requires more complex models such as the Bartlett–Lewis and Neyman–Scott models described in Chapter Seven. And when the interval of rainfall aggregation increases to a day, the Stern–Coe models of Chapter Three provide a means of using series of daily rainfall to answer practical questions about when to plant crops in semi-arid regions, and how to calculate the probability of an unbroken sequence of N days without rain after shoots have emerged. Furthermore, the unpredictability of rain – whether at the scale of individual drops or when they are aggregated over time – and the unpredictable way that soil and vegetation surfaces are wetted and what happens after they are wetted, are basic to very many problems of the natural environment, since the land phase of the hydrological cycle provides the transport mechanism for erosion of soil, sedimentation of river channels and reservoirs, transport and dispersion of pollutants in soil, groundwater and river channels. And where irregularity of rainfall in time and space is such that soil moisture is inadequate to maintain plant growth, crops, livestock, and often humans also, die.

1.2 UNCERTAINTY AND DETERMINISM

Uncertainties in rainfall occurrence are compounded where it is necessary to describe how vegetation or soil surfaces respond to rainfall, since such surfaces are irregular in orientation, composition and rigidity. Whilst, at the scale of the drainage basin, it may be possible to "average out" small-scale irregularities so that vegetation (for example) may be regarded as a single surface of large extent, other uncertainties come into force: namely the uncertainties in initial and boundary conditions which make it difficult to describe accurately how the basin will respond to rainfall, even when the rain is assumed to fall uniformly in time and uniformly over the basin's spatial extent. Whenever prediction (of flow, for example) is compared with observation, differences or "errors" are found which are the consequences of imperfections in the equations used to describe how water moves though the basin, in the specification of boundary conditions, and in the knowledge of how and where the rain really fell. Thus if a response variable, say flow denoted by q_t, is observed at equal intervals of time $t = 0, 1, 2, \ldots$, then a time-series of errors $e_t = q_t - \hat{q}_t$ results, study of which may possibly lead to modifications in model structure. Even after model modification, errors e_t will remain which are often best regarded as random variables: random in the sense that no specific causative mechanism is available to explain them. The world is too complex, and life too short, to attempt further structural changes to the model, in the forlorn hope of reducing all such errors to zero.

The complexity of factors governing how the world behaves, and the recognition of randomness in the occurrence of events, is often frustrating to physical scientists nurtured on differential equations applied to data from laboratory experiments conducted under highly controlled conditions. Biologists, and in particular ecologists taking the holistic view ("lumpers" as opposed to "splitters") have long recognised that interrelationships between factors influencing the outcomes of events, not only in the living world. Despite recent theories of chaos, this recognition is clear, for example, in the works of Darwin (1836, The voyage of the Beagle), who speculated on the relationship between atmospheric pressure, earthquakes and rainfall as follows:

> In the cases of great volcanic eruptions, as that of Coseguina, where torrents of rain fell at a time most unusual for it, and "almost unprecedented in Central America", it is not difficult to understand that the volumes of vapour and clouds of ashes might have disturbed the atmospheric equilibrium. Humboldt extends this view to the case of earthquakes without volcanic eruptions; but I can hardly conceive it possible, that the small quantities of aëriform fluids which then escape from the fissured ground, can produce such remarkable effects. There appears to be much probability in the view . . . that when the barometer is low, and when rain might naturally be expected to fall, the diminished pressure of the atmosphere over a wide extent of country, might well determine the precise day on which the earth, already stretched to the utmost by the subterranean forces, should yield, crack, and consequently tremble. It is, however, doubtful how far this idea will explain the circumstance of torrents of rain falling in the dry season during several days, after an earthquake unaccompanied by an eruption; such cases seem to bespeak some more intimate connexion between the atmospheric and subterranean regions.

An excellent discussion of causality, determinism and randomness is given by Klemes (1978), whose comments are a model of clarity. He also gives a particularly comprehensive review of relevant literature; although his paper refers specifically to hydrological analyis, it is also highly relevant to other areas of water science.

1.3 MODELS WITH RANDOM VARIABLES: TIME-SERIES MODELS

Recognition of the inherent randomness in the driving processes of the hydrological cycle, and of the fact that measured values need to be regarded as observed values of variables that fluctuate randomly, allows two broad approaches to the description of physical processes – two classes of **model** – to be identified that are useful for making predictions about future behaviour of measured variables in the physical water sciences. The first class is the class of **time-series models**, which attempt to describe, commonly in empirical terms, the statistical characteristics of a variable, or of the relationships between two or more variables, when these are observed at equal time intervals. Time-series models commonly relate one variable (possibly vector-valued) on the left-hand side of an equation to others occurring on the right-hand side. For example,

$$q_t = f(P_t, P_{t-1}, \ldots P_{t-k}; \boldsymbol{\theta}) + \varepsilon_t \tag{1.1}$$

where $f(.)$ is a function of some specified form, $\boldsymbol{\theta}$ is a vector of usually unknown parameters to be estimated from data, ε_t an error, regarded as a random variable and the variables q_t, P_t (which might be runoff from a drainage basin, and mean areal rainfall, indexed by the time subscript t, ($t = 1, 2, 3, \ldots$). The simplest form has the ε_t independently and identically distributed with zero mean and constant variance σ^2; but more generally it will be necessary to allow them to be correlated and modelled according to a form such as

$$\varepsilon_t + \beta_1\varepsilon_{t-1} + \ldots + \beta_p\varepsilon_{t-p} = a_t + \delta_1 a_{t-1} + \ldots \delta_q a_{t-q} \tag{1.1a}$$

where the a_t, instead of the ε_t, are now independently and identically distributed random variables with zero mean, constant variance σ_a^2, and the values of $p, q, \beta_i (i = 1, 2 \ldots p), \delta_i (i = 1, 2 \ldots q)$ and σ_a^2 are to be determined from data.

If, in (1.1), the $P_{t-i} (i = 0, 1 \ldots k)$ are "effective" rainfalls (i.e. rainfall minus infiltration: the component contributing to rapid runoff over a surface) and q_t is "stormflow" (streamflow minus "baseflow"); and, furthermore, if $f(.)$ is linear in the P_{t-i} and $\boldsymbol{\theta}$ of the form

$$q_t = \theta_0 P_t + \theta_1 P_{t-1} \ldots \theta_k P_{t-k} + \varepsilon_t \tag{1.2}$$

then the estimates of the $\boldsymbol{\theta}$ parameters give the ordinates of the now classical unit hydrograph. Note that if the probability distribution of q_t is conditional on the observed (and therefore fixed) values of the P_{t-i}, estimating the $\boldsymbol{\theta}$ parameters is a least-squares or multiple regression problem; but if the P_{t-i} are also random variables, as well as the q_t and ε_t, (1.2) becomes of the "transfer-function" type discussed in the classic text by Box and Jenkins (1970) which require more sophisticated procedures to estimate the $\boldsymbol{\theta}$ parameters (as well as the additional parameters $\beta_i (i = 1, 2 \ldots p), \delta_i (i = 1, 2 \ldots q)$ and σ_a^2 when the ε_t must be modelled as in (1.1a)). However, the function $f(.)$ on the left-hand side of (1.1) is very commonly non-linear in the parameters $\boldsymbol{\theta}$; each of the "lumped" or spatially averaged hydrological models relating runoff q_t to rainfall P_t, and to other variables of atmospheric condition, is an example. Tong (1990) describes some general theory for non-linear time-series models, although the non-linearities of his hydrological examples are of rather simple form.

The time-series models in this first group are useful for making short-term forecasts: the unit hydrograph model of (1.2), for example, can be used to predict the stormflow, or "rapid runoff", from a saturated river-basin, following a hypothetical rainfall sequence $P_t, P_{t-1}, \ldots$. Models of the Autoregressive Integrated Moving Average (ARIMA) form

$$q_t = \theta_0 q_{t-1} + \theta_1 q_{t-2} \ldots \theta_k q_{t-k} + \varepsilon_t + \nu_1 \varepsilon_{t-1} + \ldots \tag{1.3}$$

are used both for short-term forecasting, and for simulating streamflow sequences where, say, the frequency of failure of a complex water resource system is to be estimated. Lumped rainfall–runoff models, where $f(.)$ is non-linear and often highly so, may be used to make forecasts of streamflow a few hours or days ahead, or to "hind-cast" in the sense of estimating what the values of streamflow would have been, where the rainfall record is longer than the streamflow record. However, to test hypotheses about model structure and to give some idea about the precision of forecasts, it is necessary to postulate some specific form for the probability distribution of the errors ε_t. The classic Box–Jenkins models of ARIMA and transfer function type, for example, require that the ε_t be normally and independently distributed with zero mean and constant variance, or can be made so by some transformation of the observed variables; in the hydrological context, at least, such requirements will very rarely be met. It is possible to extend the generalised linear model (GLM) methods outlined in Chapter Two to allow the ε_t to have distributions other than the normal but belonging to the exponential family (of which the normal is a special case), but the requirement that the ε_t be statistically independent is still a source of difficulty – apart from the fact that the log likelihood function does not necessarily have a unique maximum when $f(.)$ is non-linear.

There is now an extensive literature on the use of time-series models in the water sciences, not just for rainfall–runoff modelling but also including the joint modelling of flow and sediment concentration: see, for example Hipel and McLeod (1994). For that reason, they will not be treated in this book. Despite their limitations and difficulties, time-series models are a tool of immense practical value for the water scientist, biological as well as physical.

1.4 MODELS WITH RANDOM VARIABLES: STOCHASTIC MODELS

The second group of models can be broadly defined as those expressed in terms of the probabilities that the random variables take particular values or lie within certain specified ranges. Using the appropriate terminology, this second group of models is expressed in terms of the probability that the random variable is in a particular **state**. In fact the distinction between the two groups of models – time-series models and stochastic models – being made here is more a convenience than indicative of a clear-cut difference; often, stochastic models can be expressed in time-series form. (For example, the unrestricted random walk $X_t = X_{t-1} + Z_t$ can be regarded as a time-series model (Box and Jenkins, 1970), as can the shot-noise model outlined in Chapter Six). However, time-series methods are of limited use where, as in Chapter Four, a random walk is constrained to lie between fixed boundaries.

In this second group of models, **stochastic process models**, we have a set of random variables, usually denoted by $\{X_n\}$ or $\{Z_n\}$, where n indicates discrete time (commonly $n = 0, 1, 2, \ldots$) or by $\{X(t)\}$ or $\{Z(t)\}$ when time t (for example $t \geq 0$) is continuous. For the present, we use $\{X_n\}$ and $\{X(t)\}$. The term **state space** is used to describe the range of possible values that each X_n or $X(t)$ can assume; the state space may be either discrete, where X_n or $X(t)$ take discrete values, or continuous, where X_n or $X(t)$ can take any value within some continuum. Following Cox and Miller (1965), four cases can then be distinguished when X_n or $X(t)$ develop in the single dimension of time:

1. **Discrete time, discrete state space:** here $n = \ldots -1, 0, 1, 2 \ldots$ and each X_n can take any of the discrete numerical values $x_0, x_1, \ldots x_p$, where p may be infinite. One example is given in Chapter Three, where a state variable X_n can take the values $X_n = 0$ or $X_n = 1$, according to whether day n is dry or wet.
2. **Discrete time, continuous state space:** here $n = \ldots -1, 0, 1, 2 \ldots$ and each X_n can take any value within the continuous range $L \leq X(t) \leq U$, where L, U may be $-\infty, \infty$. An example is given in Chapter Five, where $X(t)$ is the water stored in a soil moisture reservoir of maximum capacity s; at any time n, the state variable $X(t)$ may take any value in $(0, s)$.
3. **Continuous time, discrete state space:** examples are given in Chapter Six, where the state variable X_n is the number of events (floods, el Niño events, raindrop arrivals, . . .) that have occurred in the interval $(0, t)$. X_n may take any of the discrete states $0, 1, 2, \ldots$.
4. **Continuous time, continuous state space:** an example is given in Chapter Seven, where a particle (say of tracer, or contaminant) moves by a series of infinitely small positive or negative jumps, under a general directional drift. Here the state variable $X(t)$ denotes the position of the particle on the x-axis at time t.

The two groups of models (time-series models; stochastic models) are often complementary in the sense that problems can often be approached either by the use of time-series models, or by using stochastic models. We have already mentioned the kind of problem for which time-series models are used (forecasting streamflow in rivers; extending streamflow records by using longer rainfall records; simulation of complex water resource systems); in the next section we consider some kinds of problem for which stochastic models are used.

1.5 SOME APPLICATIONS OF STOCHASTIC MODELS

We may list some of the applications of stochastic models as follows.

1. One important application is to establish the **probability distribution of X_n or $X(t)$**. In discrete state-space and discrete time, for example, we may wish to obtain the probability that day n will be dry (state variable $X_n = 0$) or wet (state variable $X_n = 1$) given the state at day zero. As another example, we may wish to find the probability distribution that a sediment particle moving along a river channel is found within some small interval $(x, x + \Delta x)$ at time t. In certain cases, it is not too difficult to find algebraic expressions for the required probability distributions, but in more complex cases, it may be possible to establish only the mean value $\mathrm{E}[X_n]$ and variance, $var[X_n]$, of the state variable. And always, computer simulation of the stochastic process is a useful tool, not only to obtain an idea of the probability distribution in the more complex cases, but also to obtain a feel for how the system behaves.
2. Another important class of problem concerns **first passage times**. Given that a system (such as a reservoir with maximum capacity C) is now empty, how long will it take to fill, given the statistical characteristics of the inflows and the nature of withdrawals? If it is currently full, how long will it take to empty? Stated thus, the description of the problem is not exact; if the reservoir is currently empty and the inflows have some random structure, the time to fill is also a random variable having a probability distribution. So the problem is more exactly specified as follows: given that a system (such as a reservoir with maximum capacity C) is now empty, what is the probability distribution of the time T taken for it to become full for the first time? Or, the more limited question, what is the expected value of the time T?

   Note that the problem of determining the probability distribution of the time taken for a reservoir to become full can also be approached using time-series models, for which it would be necessary (a) to obtain a time-series model (perhaps of the form (1.3), and almost certainly incorporating some representation of the seasonal variation in inflows) of the increments to be added to the water retained in the reservoir; (b) simulate a large number of **realisations** of inflow sequences using a computer; (c) compute, for each of the realisations, the time taken for the reservoir to fill; (d) hence obtaining a histogram of the times to fill which will approximate the required probability distribution.
3. A related set of problems concerns the **distribution of lengths of time spent**

in a particular state. For example, with two discrete states $X_n = 0$ and $X_n = 1$ representing the non-occurrence or occurrence of rain on day n, we are interested in the probability distribution of the numbers of consecutive dry days. Another example is where the probability distribution is needed of the lengths of time for which a reservoir remains empty, once it has entered that state.

1.6 STATE VARIABLES VARYING IN BOTH SPACE AND TIME

The idea of a state variable varying with time is simply modified to the case where the state variable varies in a space of one dimension. One example might be where X_n is a state variable giving the moisture volume fraction of soil at 10 cm depth along a transect of equidistant points, at some fixed point in time. The idea can be extended still further to the case where the state variable varies in a space of two or three dimensions: thus X_n might be the moisture volume fraction at 10 cm depth at points in a two-dimensional space, again at some fixed point in time, observed at points which might be the nodes of a regular grid, or at irregular points in the two-dimensional space.

There is now a very large literature on the analysis of such data (Cressie, 1993) and on models appropriate for their representation; the field is commonly termed **geostatistics**, and in this book the area is touched on only very briefly in the final chapter where we discuss the computer simulation of realisations from such processes.

It should be noted that methods of geostatistical analysis cannot be applied to a particular kind of data arising from the following kind of stochastic process. Consider the case where an event may or may not occur in elements $\Delta S = \Delta x.\Delta y$ of a two-dimensional space with coordinates (x, y); suppose that a state variable $X^{(1)}$ takes values 1 or 0 according to whether the event occurs or does not occur in the element ΔS and that, if an event does occur there, another state variable $X^{(2)}$ assumes a value. For example, $X^{(1)}$ might be the presence or absence of a particular algal species in a quadrat of reservoir surface, and $X^{(2)}$ might be the number of individuals in the colony, if one is found there. If the values of $X^{(2)}$ are plotted on a graph with coordinates (x, y), the values resemble the kind of data analysable by geostatistical methods: in fact they are quite different, being a realisation of a **marked stochastic process** in two dimensions. This example is the two-dimensional analogue of the case of daily rainfall in one dimension (time), where one state variable (say $X_n^{(1)}$) takes values 1 or 0 according to the presence or absence of rain on day n, whilst the second state variable $X_n^{(2)}$ takes positive values representing the depth of rain when day n is wet.

Examples given in the present section have so far considered that time was fixed, with state variables varying over two-dimensional space. By allowing a state variable to vary in time as well as spatially, the general **spatial–temporal process** is obtained; where the temporal process is a point process (Chapter Six) the treatment of such processes is given in Cox and Isham (1980) and will not be dealt with in this book.

1.7 CONTENTS OF FOLLOWING CHAPTERS

The following Chapter Two gives an outline of some of the mathematical and statistical procedures used in the subsequent chapters. Chapter Three discusses the use of Markov chains for the analysis of daily rainfall data, based on the work of Stern and Coe (1984); an example is given using daily rainfall data from Manaus, Amazonia. Chapter Four extends the use of Markov chains to the modelling of reservoir behaviour, while Chapter Five uses Markov chain theory to the modelling of rainfall–runoff relationships in drainage basins. Chapter Six deals with Poisson processes and their generalisation renewal processes, which have been widely used in many areas of water science; Chapter Seven develops the simple random walk to the modelling of diffusion processes and the modelling of tracer movement. The concluding Chapter Eight gives a summary account of some computer simulation methods which are an important tool for understanding the nature of stochastic processes, whether in time or space.

Chapter Two

Mathematical preliminaries

2.1 THE NEED

Common to many areas of water science is the notion of "variability" amongst observations or measurements repeated in time or space. Information contained in a set of such measurements (**data**) can be summarised in terms of **statistics** describing, in terms of a small set of numbers, the position, spread and shape of histograms obtained from the data. Calculation of such descriptive statistics is a necessary first step towards understanding any natural phenomenon, but sooner or later it becomes necessary to formulate and test hypotheses concerning the mechanisms giving rise to the data, and to make predictions based on current understanding. The processes of formulating a description of the mechanisms, using it to make predictions, comparing prediction with observation, and modifying the description where necessary, are collectively called **modelling**, and the description of the mechanism, if expressed in mathematical terminology, is called a **mathematical model**. Testing hypotheses about mathematical models, and deciding whether an observation is so different from what a model predicts as to necessitate some rethinking about its structure, requires methods based in **probability theory**.

The sections that follow give only a very brief account of basic probability theory, and the reader is referred to fuller texts on the subject for further reading. The book by Feller (1957) is a classic work that is full of interesting examples which well repay study; a more modern work is that of Grimmett and Stirzaker (1992).

2.1 SAMPLE SPACES, EVENTS AND PROBABILITY

2.1.1 Sample spaces

The foundations on which the edifice of probability theory is constructed are the concepts of (1) sample space, (2) events defined on a sample space and

(3) probability or measure associated with each event. As an example, the occurrence of the climatological event "el Niño" is subject to uncertainty, being a consequence of the interaction of many imperfectly understood factors; if we were to consider the occurrences of el Niño events in the next four years, there are $2^4 = 16$ possible outcomes, which can be represented by the set {0000, 0001, 0010, 0011, 0100, 0101, 0110, 0111, 1000, 1001, 1010, 1011, 1100, 1101, 1110 and 1111}, where 1 and 0 represent the occurrence or non-occurrence of an el Niño event in a particular year. Thus 0000 denotes the non-occurrence of el Niño in any of the four years, while 1111 represents the (much more unlikely) occurrence of an el Niño event in all four years. The set enclosed within the curly brackets {.} is an example of a **sample space**, commonly denoted by the symbol $\mathbf{\Omega}$, and entities like 0000, 0001 are **points** in this sample space.

2.1.2 Events

Events are subsets of points within $\mathbf{\Omega}$: thus the event "only one el Niño event occurs in the four years" consists of the points 0001, 0010, 0100, 1000; and the event "one or more el Niño events occur in the four years" contains the set of 15 points obtained by omitting the point 0000 from $\mathbf{\Omega}$. The point 0000 in $\mathbf{\Omega}$ is obviously the event "no el Niño event occurs in the four years"; such events, consisting of single points of $\mathbf{\Omega}$, are termed **elementary events**. To any event A defined in $\mathbf{\Omega}$ there is a complementary event, A^c, containing all elementary events not lying in $\mathbf{\Omega}$. Thus the events A: "no el Niño in four years" is complementary to A^c : "at least one el Niño in four years".

Given two or more events denoted by $A, B, \ldots$, further events may be defined, using the union and intersection symbols $\cup$ and $\cap$. Thus if A, B are the events A: "two el Niño events or more in the four years" and B: "no el Niño events in consecutive years", the event $A \cup B$ (read as "A union B") is the event containing all elementary events occurring either in A or in B. The event A contains the 11 elementary events {0011, 0101, 0110, 0111, 1001, 1010, 1011, 1100, 1101, 1110, 1111} and the event B contains the eight elementary events {0000, 0001, 0010, 0100, 0101, 1000, 1001, 1010}; therefore the event $A \cup B$ is the event containing all 16 elementary events, i.e. $A \cup B = \mathbf{\Omega}$. The event $A \cap B$, read as "A intersection B", contains the elementary events which are common to both A and B; in the example, $A \cap B = \{0101, 1001, 1010\}$. Where two events A, B have no elementary events in common, we write $A \cap B = \varnothing$, the empty set, and call $\varnothing$ the **impossible event**. The events A and B are then said to be **mutually exclusive** or **disjoint.** The event $\mathbf{\Omega}$ containing all elementary events is called the **certain event,** since one of the elementary events of which it is constituted must occur. Thus $A \cup A^c = \mathbf{\Omega}$, since either A or A^c must occur, and $A \cap A^c = \varnothing$, since if A occurs, A^c does not. Whilst all events are subsets of $\mathbf{\Omega}$, there are mathematical reasons why it is not necessarily true that all the subsets of $\mathbf{\Omega}$ are events; and for practical purposes, we can think of events as being a kind of subcollection of all the possible subsets in $\mathbf{\Omega}$. This subcollection, often denoted by F, has the following properties:

1. If events A and $B \in F$ ("are in F"), then $A \cup B \in F$ and $A \cap B \in F$;
2. If $A \in F$ then $A^c \in F$;
3. The empty set $\varnothing$ belongs to F.

Here F is called a **field.** The el Niño example used above to introduce the idea of sample space contains a finite set of elementary events. More generally, a sample space can contain an infinite number of ordered elementary events: consider the sample space with elementary events $A_1, A_2, A_3 \ldots$ where A_i: "the next occurrence of el Niño occurs is in year i, counting year 0 as the present". It can be shown that the collection F can be generalised to accommodate such cases, and extended still further where the "points" in the sample space form a continuum; the mathematical details are important in that they give a rigour to the basis of probability theory, but are unimportant for the purposes of this book.

An example used above concerned the occurrence of el Niño events during a four-year period. In practice, the periods of interest are likely to be very much longer than four years; Hamilton and Garcia (1986) presented a chronological sequence of strong el Niño events in Peru, reconstructed from historical data, for the period 1525–1982, showing that events occurred in the years 1541, 1578, 1614, 1624, 1652, 1701, 1720, 1728, 1747, 1763, 1770, 1791, 1804, 1814, 1828, 1891, 1925 and 1982. These data are shown graphically in the upper half of Figure 2.1; the lower half of the figure, shown to illustrate the difficulty of reconstructing such series, shows another sequence of el Niño events through the same period and almost the same region (western South America) reconstructed by Quinn et al. (1978). Here, we use only the data of Hamilton and Garcia (1986), where we see that two events occurred in the (incomplete) sixteenth century, three in the seventeenth, seven in the eighteenth, four in the nineteenth, and two in the (incomplete) twentieth century. Given this record of el Niño occurrences for the period 1525–1982, climatologists and hydrologists will typically pose questions like: "Have el Niño events occurred 'at random' throughout the period of reconstruction, or have they occurred in 'clumps'? Is there evidence that the frequency of occurrence varied throughout the period of reconstruction? If the frequency of occurrence varies, is there evidence of increasing or decreasing trend in the occurrence of el Niño events?"

To answer such questions, we need to use the tools of (1) probability theory, which allows us to estimate the probability of an el Niño event; (2) statistical modelling, which allow us to evaluate the evidence for and against the hypothesis of random occurrences, for and against the hypothesis of trends.

2.1.3 Probability

We wish to give a quantitative measure to the likelihood that an event occurs. Going back to the el Niño example, we can count the number of years N_e in

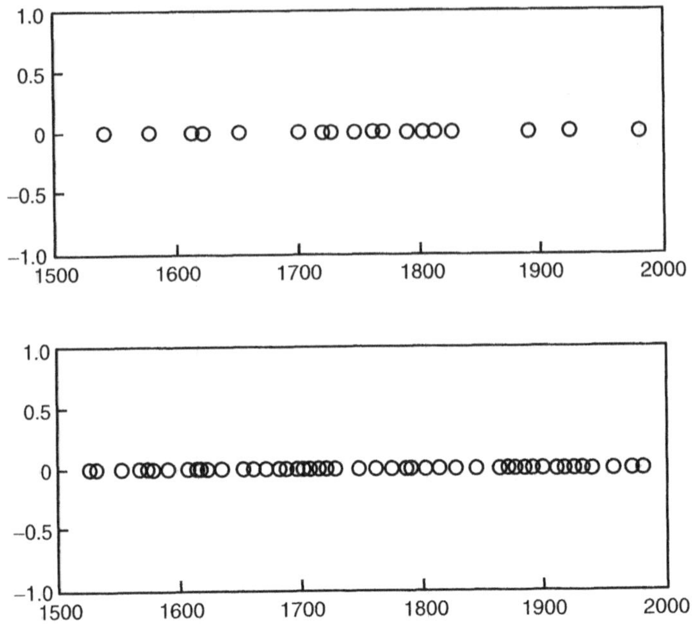

Figure 2.1 Plot of historic el Niño events in Peru, 1525–82 (top) reconstructed by Hamilton and Garcia (1986): plot of historic el Niño events in western South America (bottom) by Quinn et al. (1978)

which el Niño occurred in a record of N years; then $p(N_e) = N_e/N$ will provide a measure of how often el Niño occurs in an arbitrary year. As the length N of record increases, then intuitively the proportion $p(.)$ will tend towards some more or less stable value, π say (we assume, unrealistically, that climate is "stationary"). Clearly the proportion p must lie between 0 and 1, and we can think of

$$\pi = \lim_{N \to \infty} p(N_e)$$

as the **probability** of the event "el Niño occurs in an arbitrary year". There are serious philosophical weaknesses in this superficial argument, but it serves the purposes of this book if, corresponding to every event in F defined on the sample space $\mathbf{\Omega}$, we define a measure or **probability,** denoted by $P[A]$, having the following properties:

(a) $$0 \leq P[A] \leq 1; \tag{2.1a}$$

(b) $$P[\varnothing] = 0 \quad \text{and} \quad P[\mathbf{\Omega}] = 1; \tag{2.1b}$$

(c) if A and B are mutually exclusive (disjoint) events, then

$$P[A \cup B] = P[A] + P[B]. \tag{2.1c}$$

Standard results derived from these axioms (and easily verified using a Venn diagram) include the following:

(i)
$$P[A^c] = 1 - P[A] \tag{2.2}$$

since the events A and A^c are mutually exclusive:

$$P[A] + P[A^c] = P[A \cup A^c] = P[\Omega] = 1.$$

(ii) For any two events A, B in F,

$$P[A \cup B] = P[A] + P[B] - P[A \cap B]. \tag{2.3}$$

More generally,

$$P[A \cup B \cup C] = P[A] + P[B] + P[C] - P[A \cap B] - P[B \cap C] - P[C \cap A] + P[A \cap B \cap C],$$

and

$$P[A_1 \cup A_2 \cup A_3 \ldots \cup A_n] = \sum_i P[A_i] - \sum_{i<j} P[A_i \cap A_j] + \sum_{i<j<k} P[A_i \cap A_j \cap A_k] - \ldots + (-1)^{n+1} P[A_1 \cap A_2 \cap A_3 \cap \ldots \cap A_n]. \tag{2.4}$$

2.1.4 Conditional probability

It is frequently necessary to make probability statements about an event A, conditional on the occurrence of some other event B. Suppose, for example, that B is the event B: "Pacific sea surface temperature exceeds 25 °C", whilst A: "an el Niño event occurs". Assuming that records exist of both sea surface temperatures in the Pacific Ocean and the years in which el Niño occurred, suppose that sea surface temperatures exceeded 25 °C in N_t years, and that in N_e of these an el Niño event occurred. Then intuition suggests that the conditional event, $A \mid B$ should be given a probability estimated by N_e/N_t, or $(N_e/N)/(N_t/N)$ where N is the total number of years of record. But N_e/N is the proportion of years in which both A and B (namely, the event $A \cap B$) occurred together, and N_t/N is the proportion of years in which B occurred. This suggests the formal definition of the conditional probability $P[A \mid B]$: **Provided $P[B] > 0$, the conditional probability of the event A, given that event B occurs, is**

$$P[A \mid B] = P[A \cap B]/P[B]. \tag{2.5}$$

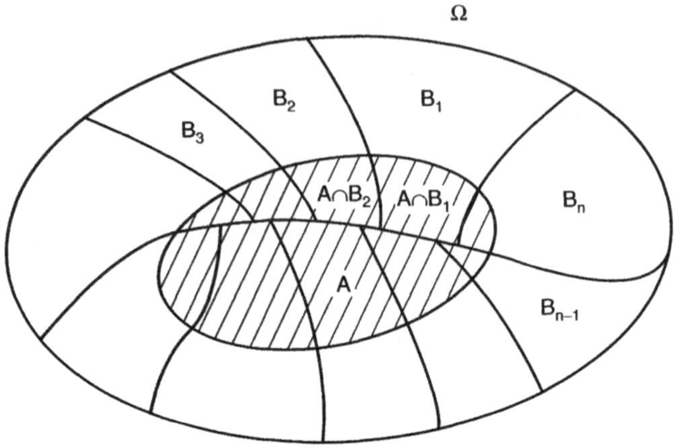

Figure 2.2 Diagram showing: Ω, the sample space; A, an event within it; subdivision of Ω into n (possibly infinite) mutually exclusive events B_1, B_2, B_3 . . . B_n such that $B_1 \cup B_2 \cup B_3 \ldots \cup B_n = \Omega$; representation of A in terms of the union of events $A \cap B_1$, $A \cap B_2, \ldots A \cap B_n$.

2.1.5 The theorem of total probability

The definition of conditional probability, and the definition of mutually exclusive events given earlier, combine to give a theorem – the theorem of total probability – which is crucially important for very many applications of probability theory. Formal mathematical proof of this and the following material is omitted, but the theorem can be verified by reference to the Venn diagram shown in Figure 2.2.

Suppose that the events $B_1, B_2, B_3, \ldots B_n$ are a partition of the sample space **Ω**: that is, the events B_i are mutually exclusive ($B_i \cap B_j = \varnothing$ for all $i \neq j$) and their union is the entire sample space ($B_1 \cup B_2 \cup B_3 \ldots \cup B_n = \mathbf{\Omega}$). Then the theorem of total probability states that

$$P[A] = P[A \mid B_1].P[B_1] + P[A \mid B_2].P[B_2] + \ldots + P[A \mid B_n].P[B_n]$$

or

$$P[A] = \sum_{i=1}^{n} P[A \mid B_i].P[B_i]. \tag{2.6}$$

2.1.6 Independence

In general, occurrence of an event B will influence the outcome of an event A, the conditional probability $P[A \mid B]$ replacing the original probability $P[A]$. If, however, the occurrence of the event B has no influence on the outcome of the event A, the events A and B are **independent**. Thus independence implies that

$$P[A \mid B] = P[A]. \tag{2.7}$$

But, by the definition of conditional probability (2.5), we have, when $P[B] > 0$,

$$P[A \cap B]/P[B] = P[A]$$

giving

$$P[A \cap B] = P[A].P[B]. \tag{2.8}$$

More generally, events $A_1, A_2, A_3 \ldots A_n$ are independent if

$$P[A_i \cap A_j \cap A_k \ldots \cap A_p] = P[A_i].P[A_j].P[A_k] \ldots P[A_p]$$

where $A_i, A_j, A_k, \ldots A_p$ are any subset of the events $A_1, A_2, A_3 \ldots A_n$.

Note that independence, which is a statement about the probabilities of events, and mutual exclusion, which is a statement about events in relation to their common sample space, are two different concepts. To illustrate the distinction, suppose that $\{i, j\}$, where i, j values $= 1, 2, \ldots 6$ represent the scores obtained with two successive throws of a fair dice; if A is the event A: "first score better than four" and B is the event B: "second score better than five", then A and B and are independent but not mutually exclusive, as may be seen by enumerating the 36 points i, j in the sample space which, by virtue of the fairness of the dice, are all equally probable with probability 1/36 . Similarly, the events C: "first score better than four" and D: "total of two throws equal to three or less" are mutually exclusive ($P[C \cap D] = \varnothing$) but not independent, since

$$P[C \cap D] \neq P[C].P[D].$$

A further instructive example, given by Feller, concerns a family of three children, with sample space denoted by BBB, BBG, BGB, BGG, GBB, GBG, GGB, GGG. If all eight elementary events have equal probability $\frac{1}{8}$, the events A: "the family has children of both sexes" and B: "the family has one daughter or less" have probabilities $\frac{3}{4}$ and $\frac{1}{2}$ respectively, and $P[A \cap B] = \frac{3}{8}$; since

$$P[A \cap B] = P[A].P[B],$$

the events A and B are independent. It is not true, however, that A and B are independent for families of two or four children, as may be verified by enumerating the points of the sample space.

2.1.7 Random variables

Usually there is less interest in the sample space and events that occur in it, than in some function derived from the events. In the example of occurrences of el Niño events in four years – or more usefully, in N years – a climatologist will commonly be more interested in the probabilities of occurrence of zero el Niño events, one el Niño event, two el Niño events, . . . (irrespective of whether the one el Niño event occurred in the first of the N years, or the second of the N years, . . .). In practice, therefore, we look for a function which effects a mapping from the 2^N points in the sample space $\mathbf{\Omega}$ into the points $0, 1, 2, \ldots N$ of the real line. The point $0000\ldots0$ (consisting of N zeros) in $\mathbf{\Omega}$ will map into the point $x = 0$ of this line; there will be N points – namely, $1000\ldots0$; $0100\ldots0$; $0000\ldots1$ – which map into the point $x = 1$; and so on. Since each elementary event in $\mathbf{\Omega}$ has a probability, the mapping of an elementary event into (say) $x = r$ also represents a transfer of probability to that point on the real line; and having mapped each and every elementary event from $\mathbf{\Omega}$, we can speak of a random variable X, assuming the values $0, 1, 2, \ldots N$ with probabilities $p_0, p_1, p_2, \ldots$ such that the sum $p_0 + p_1 + p_2 + \ldots + p_N = 1$. In symbols, we can write the **probability density function (pdf)** of the **random variable** X as

$$P[X = x] = p_x. \tag{2.9}$$

The example used to illustrate the mapping process took the points of $\mathbf{\Omega}$ into the points $0, 1, 2, \ldots N$, but the mathematical basis permits the number of points N on the real line to be infinite. The mathematics also shows that there are particular advantages in working with the distribution function $F(x)$, defined as follows: **The distribution function of a random variable X is the function $F(x)$, with value between 0 and 1, such that**

$$F(x) = P[X \leq x] \tag{2.10}$$

and for the el Niño example, we have from (2.9)

$$F(x) = p_0 + p_1 + p_2 + \ldots + p_x.$$

In many of the practical applications dealt with in this book, the random variable X can in theory take on continuous values; sea surface temperature in the Pacific Ocean is one example (although practical limitations of instrument design may result in field measurements that can increase or decrease only by a multiple of some increment), and where we work with continuous variables we assume that the distribution function $F(x)$ is differentiable with $\mathrm{d}F(x)/\mathrm{d}x = f(x)$, the pdf of the continuous random variable X. Thus

$$F(x) = \int_{-\infty}^{x} f(u)\mathrm{d}u \tag{2.11}$$

where $f(.)$ is the pdf of the continuous random variable X.

Hitherto the discussion of independence, distribution functions and pdfs have been limited to one random variable X; however, we are very often concerned with the joint characteristics of two or more random variables, $X, Y \ldots$. Two discrete random variables X and Y are said to be independent if the events $[X = x]$ and $[Y = y]$ are independent for all x and y. Two continuous random variables X and Y are independent if the events $[X \leq x]$ and $[Y \leq y]$ are independent for all x and y. The joint distribution function $F(x, y)$ of two random variables X and Y is

$$F(x, y) = P[(X \leq x) \cap (Y \leq y)].$$

The (joint) pdf of X and Y is $f(x, y)$, where

$$F(x, y) = \int_{-\infty}^{x} \int_{-\infty}^{y} f(u, v)\, \mathrm{d}u\, \mathrm{d}v$$

If the partial derivatives exist, $f(x, y) = \partial^2 F(x, y)/\partial x \partial y$.

2.1.8 Expected values of random variables and functions of random variables

In accordance with intuition, the "expected" value of a discrete random variable X, denoted by $\mathrm{E}[X]$, is a weighted sum of its possible values, the weights being probabilities. In symbols,

$$\mathrm{E}[X] = \sum_{x} X P[X = x].$$

The expected value of a continuous random variable X with density function $f(x)$ is

$$\mathrm{E}[X] = \int_{-\infty}^{\infty} x f(x) \mathrm{d}x.$$

The expected value of the pdf $f(x)$ gives its mean value, denoted by μ.

The concept of the "expected value of a random variable X" can be extended to include the definition of expected values of functions of random variables: if $g(X)$ is a function of the random variable X, then its expected value is defined as

$$\mathrm{E}[g(X)] = \sum_{x} g(x) . P[X = x]$$

in the discrete case, and

$$\mathrm{E}[g(X)] = \int_{-\infty}^{\infty} g(x) f(x) \mathrm{d}x$$

in the continuous case. Of particular importance is $g(x) = (x - \mu)^2$, for then

$E[(X-\mu)^2]$ gives the **variance** of X, denoted by σ^2, measuring the dispersion of the pdf $f(x)$. Some simple algebra gives the useful relation

$$\begin{aligned}\operatorname{var}[X] &= E[(X-\mu)^2] \\ &= E[X^2] - E[X]^2.\end{aligned}$$

Higher **moments about the mean** of the random variable X are obtained when $g(x) = (x-\mu)^k$ for $k > 2$. When $k = 3$,

$$E[(X-\mu)^3]/\sigma^3$$

is the coefficient of skewness, and when $k = 4$,

$$E[X-\mu)^4]/\sigma^4$$

is the coefficient of kurtosis, respectively measures of asymmetry of form and "peakedness" of the pdf $f(x)$.

2.1.9 Marginal and conditional distributions

The joint pdf $f(x, y)$ of two random variables X and Y can be pictured graphically in three dimensions, with X and Y plotted along the two perpendicular axes in the horizontal plane, and the density $f(x, y)$ plotted in the direction of the vertical axis. Thus $f(x, y)$ can be represented as a surface, with the ordinate $z = f(x, y)$ always positive, and with the integrated volume between the surface and the horizontal plane equal to one. An alternative way of visualizing the joint pdf $f(x, y)$ uses two dimensions, corresponding to the plane with perpendicular axes X and Y; in this alternative visualization, a unit mass is distributed over the X–Y plane, in such a way that the density at the point (x, y) is $f(x, y)$. We now extend this mental picture of $f(x, y)$ as a density of mass and consider what happens if the unit mass distributed across the X–Y plane is compressed in the Y-direction, thereby resulting in a distribution of the unit mass along the X-axis only; the result of this compression is the **marginal density function** $f_X(x)$, defined symbolically by

$$f_X(x) = \int_{-\infty}^{\infty} f(x, y)\mathrm{d}y.$$

Similarly, the marginal density function $f_Y(y)$ is

$$f_Y(y) = \int_{-\infty}^{\infty} f(x, y)\mathrm{d}x.$$

The **conditional distribution function of the random variable Y**, given that $X = x$, as

$$P[Y \leq y \mid X = x] = F_{Y|X}(y \mid x)$$

$$= \int_{-\infty}^{u=y} f(x, u)/f_X(x)\mathrm{d}u$$

and the **conditional probability density function** $f_{Y|X}(y\,|\,x)$ is defined by

$$f_{Y|X}(y\,|\,x) = f(x, y)/f_X(x).$$

We can interpret these definitions by returning again to the mental picture of a unit of mass distributed over the X–Y plane with density $f(x, y)$. If, for a fixed value of X, $X = x$ say, we take a thin strip of thickness $\mathrm{d}x$, centred on X, across the X–Y plane in the Y-direction, the total probability (mass) lying within the strip will be $f_X(x)\mathrm{d}x$; and the probability that the random variable Y, given that $X = x$, falls within any small interval $(y, y + \mathrm{d}y)$ will be proportional to $f(x, y)\mathrm{d}x\mathrm{d}y$. The constant of proportionality must be $1/f_X(x)\mathrm{d}x$, this being needed to ensure that the total probability in the strip integrates to unity, as theory requires. Using this intuitive argument, therefore, we have

$$f_{Y|X}(y\,|\,x) = f(x, y)/f_X(x),$$

corresponding to the formal definition.

We can extend the idea of expectation to the bivariate case also: the expected value of the function $g(X, Y)$ of the two random variables X and Y is

$$\mathrm{E}[g(X, Y)] = \int_{-\infty}^{\infty}\int_{-\infty}^{\infty} g(x, y)f(x, y)\mathrm{d}x\mathrm{d}y.$$

In addition, the **conditional expectation of Y, given that** $X = x$, is

$$\mathrm{E}[Y\,|\,X = x] = \int_{-\infty}^{\infty} y.f_{Y|X}(y\,|\,x)\mathrm{d}y$$

$$= \int_{-\infty}^{\infty} y.f(x, y)/f_X(x)\mathrm{d}y.$$

Writing $\mathrm{E}[Y\,|\,X = x] = \mathrm{E}[Y\,|\,x]$, it can be shown that

$$\mathrm{E}[\mathrm{E}[Y\,|\,x]] = \mathrm{E}[Y]$$

and this gives the following extremely useful result:

$$\mathrm{E}[Y] = \int_{-\infty}^{\infty} \mathrm{E}[Y\,|\,X = x]f_X(x)\mathrm{d}x.$$

In the above, we have restricted the argument to the case of two jointly distributed random variables, X and Y. Clearly the same kind of arguments can be used for the case where there are three random variables X, Y and Z, for which there are three univariate marginal pdfs $f_X(x)$, $f_Y(y)$ and $f_Z(z)$; three bivariate marginal pdfs

$$f_{XY}(x, y),\ f_{Yz}(y, z) \quad \text{and} \quad f_{ZX}(x, z);$$

three univariate conditional pdfs of the form

$$f_{X|Y,Z}(x \mid y, z) = f(x, y, z)/f_{YZ}(y, z),$$
$$f_{Y|X,Z}(y \mid x, z) = f(x, y, z)/f_{XZ}(x, z),$$
$$f_{Z|X,Y}(z \mid x, y) = f(x, y, z)/f_{XY}(x, y);$$

three bivariate conditional pdfs

$$f_{X,Y|Z}(x, y \mid z) = f(x, y, z)/f_Z(z),$$
$$f_{Y,Z|X}(y, z \mid x) = f(x, y, z)/f_X(x),$$

and $$f_{Z,X|Y}(z, x \mid y) = f(x, y, z)/f_Y(y),$$

together with a further six conditional pdfs of the form

$$f_{Y|X}(y \mid x),\ f_{X|Y}(x \mid y)$$

obtained from the marginal distributions $f_{XY}(x, y)$, $f_{YZ}(y, z)$ and $f_{ZX}(z, x)$.

The extension to the case of more than three variates is straightforward, and expressions are written more simply if vectorial notation is used: for example, if

$$\boldsymbol{x}_1 = \{x_1, x_2, x_3 \ldots x_p\} \quad \text{and} \quad \boldsymbol{x}_2 = \{x_{p+1}, x_{p+2}, x_{p+3}, \ldots x_{p+q}\},$$

then the conditional pdf of $\boldsymbol{x}_1$ given $\boldsymbol{x}_2$ can be written

$$f_{1|2}(\boldsymbol{x}_1 \mid \boldsymbol{x}_2) = f(\boldsymbol{x}_1, \boldsymbol{x}_2)/f_2(\boldsymbol{x}_2).$$

2.1.10 Important examples of pdfs

The preceding sections have dealt in a general way with some of the characteristics of pdfs; some specific examples follow that are used in later chapters.

Discrete pdfs

The **Bernoulli pdf** describes the characteristics of a discrete random variable X taking values that can be denoted by 0 or 1, typically the absence or presence of an event, or the "failure" or "success" of a single trial:

$$P[X = 0] = 1 - \pi \qquad (= f_0, \text{say})$$
$$P[X = 1] = \pi \qquad (= f_1)$$

where $0 \leq \pi \leq 1$. The expected value of X is $E[X] = 0.(1 - \pi) + 1.\pi = \pi$, and the variance of X is

$$\text{var}[X] = E[(X - E[X])^2] = E[X^2] - E[X]^2 = 1^2.\pi - \pi^2$$

or $\text{var}[X] = \pi(1 - \pi)$.

The **binomial pdf** describes the characteristics of a random variable X, the number of "successes" in a sequence of N **independent** Bernoulli trials:

$$P[X = x] = N!/[x!(N - x)!]\pi^x(1 - \pi)^{N-x},$$

where x, the number of successes, can take the values $0, 1, 2, \ldots N$. From the definition of the mean and variance in terms of expected values, it can be verified that the mean and variance of the binomial pdf are

$$E[X] = N\pi \qquad \text{var}[X] = N\pi(1 - \pi).$$

If $Y = x/N$ is a new random variable representing the proportion of successes in the N trials, Y can be shown to have mean and variance

$$E[Y] = \pi \qquad \text{var}[Y] = \pi(1 - \pi)/N.$$

The **Poisson pdf** can be regarded as a limiting form of the random variable X of a binomial pdf, when the number of trials N increases without limit, and the probability π of a success decreases to zero, such that the product $N.\pi$ tends to a finite limit, say λ. It is found that

$$P[X = x] = \lambda^x \exp(-\lambda)/x! \qquad x = 0, 1, 2, \ldots.$$

The mean and variance of the Poisson distribution are found by their definitions to be

$$E[X] = \lambda \qquad \text{var}[X] = \lambda.$$

A **Poisson process** is a phenomenon in which events occur in time such that, with X denoting the number of events,

1. $P[X = 1$ in any small time interval $\delta t]$ is proportional to $\delta t = \lambda\delta t$, say;
2. If δt_i and δt_j are any two non-overlapping time intervals, events in δt_i are independent of events in δt_j, $i \neq j$;
3. $P[X > 1$ in any small time interval $\delta t] = o(\delta t)$ and is therefore negligible.

The number of events X then occurring in any time interval of length $T = N\delta t$ then has a pdf which is the limit, as N increases indefinitely and δt tends to zero, of

$$P[X = x] = N!/[x!(N - x)!](\lambda\delta t)^x(1 - \lambda\delta t)^{N-x}$$

which can be shown to be

$$P[X = x] = (\lambda T)^x \exp(-\lambda T)/x! \qquad x = 0, 1, 2, \ldots; T > 0$$

whence the mean and variance of X are found by definition to be

$$E[X] = \lambda T \quad \text{and} \quad \text{var}[X] = \lambda T.$$

A starting point for study of the occurrence of el Niño events in the period 1525–1982 (see the data of Hamilton and Garcia, 1986, given above in section 2.1.2) might be the hypothesis that events occur according to a Poisson process, and statistical analysis of the data would seek to justify this hypothesis, or to demonstrate that it should be rejected. Furthermore, although we have discussed the Poisson process as one developing in time, we could equally well apply it to phenomena varying in a space of one dimension (such as the occurrence of a botanical species along the course of a river or sampling transect) or in a space of two dimensions (such as the occurrence of lightning flashes in a satellite image).

It was stated above that, for a Poisson process,

$$P[X = x] = (\lambda T)^x \exp(-\lambda T)/x!$$

with mean and variance of the random variable X both equal to λT. Where events occur in clusters or groups, the Poisson model will be inappropriate, since the effect of clustering will increase the dispersion of data as measured by the variance. A distribution providing a good representation of clustered events is the **negative binomial,** for which the variance σ^2 and mean μ are related by $\sigma^2 = \mu + \mu^2/k$; when the parameter k is very large, the distribution tends to Poisson form. The negative binomial distribution can be derived by assuming that the parameter λ of a Poisson process is itself distributed according to a gamma distribution described below. The pdf of the negative binomial distribution involves two parameters μ and κ, of the form

$$P[X = x] = (x + \kappa - 1)!/[(\kappa - 1)!x!]\,\{\mu/(\mu + \kappa)\}^x\{1 + \mu/\kappa\}^{-\kappa} \qquad x = 0, 1, 2, \ldots,$$

where μ is the mean of the distribution, and its variance σ^2 is related to the mean μ by $\sigma^2 = \mu + \mu^2/\kappa$.

There are other distributions which describe the distribution of events occurring in clusters. One is the **Neyman type A distribution**, obtained when the number of groups per time interval has a Poisson distribution with parameter λ_1, and the number of events per group also has a Poisson distribution with parameter λ_2. Its pdf is

$$P[X = 0] = \exp\{-\lambda_1(1 - \exp(-\lambda_2))\}$$

$$P[X = x] = (\lambda_1/x)\exp(-\lambda_2).\sum_{j=1}^{x} \lambda_2^j P[X = x - j]/(j - 1)! \qquad x = 1, 2, 3, \ldots.$$

The Neyman type A is less skew than the negative binomial, and when λ_2 tends to zero the distribution tends to the Poisson form.

The **Pólya–Aeppli distribution** also describes phenomena where groups of events occur according to a Poisson distribution with parameter λ_1, and the number of events per group has a geometric distribution with parameter π. Its pdf is given by

$$P[X = 0] = \exp(-\lambda_1)$$

$$P[X = x] = (\pi\lambda_1/x) \sum_{j=0}^{x-1} (x - j)(1 - \pi)^{x-j-1} P[X = j] \qquad x = 1, 2, 3, \ldots$$

As π tends to 0 the distribution tends to Poisson form; its mean is $\mu = \lambda/\pi$ and its variance $\sigma^2 = \lambda(2 - \pi)/\pi^2$.

The **Poisson–Pascal distribution** has three parameters κ, π_1 and λ

$$P[X = 0] = \exp[-\lambda(1 - \theta^{-\kappa})]$$

$$P[X = x] = \lambda\pi\kappa/[x\theta^{\kappa+1}] \sum_{j=1}^{\infty} \{\pi/\theta\}^{j-1}$$

$$.\{(\kappa + j - 1)!/[\kappa!(j - 1)!]\}.P[X = x - j] \qquad x = 1, 2, \ldots,$$

where $\theta = 1 + \pi$. It is the distribution obtained when the number of groups of events per unit time has a Poisson distribution, whilst the number of events per group has a negative binomial distribution. Finally, the **Poisson–log normal distribution**, with skewness greater than the negative binomial, is obtained from a sum of Poisson pdfs whose parameters are log-normally distributed (i.e. their natural logarithms have a normal distribution). The pdf of this distribution is

$$P[X = x] = 1/[x!\sigma\surd(2\pi)]. \int_{z=0}^{\infty} \exp(-z)z^{x-1}\exp[-(\log_e z - \mu)^2/(2\sigma^2)]dz.$$

The distributions presented above all describe the probabilities of the events $X = x$ during discrete time intervals. Where emphasis is on the occurrence of events in a time continuum, a different kind of model is used: two which have received particular study, and which are expounded later in Chapter Six, are **Bartlett–Lewis process** and the **Neymann–Scott process**. Both are based on a Poisson process for the occurrence of groups of events, but differ in their descriptions of the occurrences of individual events within groups.

Continuous pdfs

It can be shown that, where events occur in one dimension denoted by $\mathbf{R}^1$ (commonly time, but see earlier comment) according to a Poisson process, the interval T between successive events has a distribution function given by

$$F(t) = \mathrm{P}[T \leq t] = 1 - \exp(-\lambda t) \qquad \lambda > 0, t > 0$$

for non-negative t, since $P[T > t] = 1 - P[T \leq t]$ is the probability that no event has occurred up to time t, $\exp(-\lambda t)$. Here $F(t)$ is the distribution function of the **exponential distribution**, with pdf found by differentiation of $F(t)$ with respect to t as

$$f(t) = \lambda \exp(-\lambda t) \qquad \lambda > 0, t > 0$$

whose mean and variance are given by $\mu = 1/\lambda, \sigma^2 = 1/\lambda^2$.

The sum U of k independent random variables $U = T_1 + T_2 + T_3 \ldots + T_k$, where T_i $(i = 1 \ldots k)$ are each from an exponential distribution with the same parameter λ, is a random variable with pdf

$$f(u) = \lambda(\lambda u)^{k-1}\exp(-\lambda u)/k! \qquad k, \lambda > 0; u \geq 0$$

which is a particular case (k an integer) of the **gamma distribution** with pdf

$$f(u) = \lambda(\lambda u)^{k-1}\exp(-\lambda u)/\Gamma(k) \qquad k, \lambda > 0; u \geq 0,$$

where $\Gamma(k)$ is the (complete) gamma function defined for any non-negative number, not necessarily an integer. The mean, variance and skewness of $f(u)$ are k/λ, k/λ^2 and $2/\sqrt{k}$ respectively, the latter showing that asymmetry decreases as k increases. The parameter k determines the shape of the distribution; if $0 < k < 1$, $f(u)$ is infinite at $u = 0$ and decreases monotonically as $u \to \infty$; if $k = 1, f(u)$ reduces to the exponential pdf. If $k > 1, f(u)$ is zero at $u = 0$ and has a single maximum.

Of outstanding importance is the **normal distribution**, with pdf

$$f(x) = 1/[\sigma\sqrt{(2\pi)}]\exp\{-\tfrac{1}{2}(x - \mu)^2/\sigma^2\} \qquad -\infty < x < \infty$$

with mean and variance μ, σ^2 respectively. In addition to affording a good description of the dispersion of many observations in nature, the normal pdf – commonly denoted by $N(\mu, \sigma^2)$ – is related to all distributions (the binomial, Poisson, gamma included) derived by considering sums of independently distributed random variables with common pdf, through the **central limit theorem**, given in the next section.

A distribution describing the passage time, t, to a fixed barrier positioned at $x = a$, of a particle making a random walk with drift v along the x-axis is the **inverse normal distribution** with pdf

$$f(t) = a/[\sigma\sqrt{(2\pi t^3)}]\exp\{-(a - vt)^2/(2\sigma^2 t)\} \qquad t > 0$$

which can be reparameterised in terms of new parameters κ, ρ as

$$f(t) = ((\kappa/\rho)/2\pi t^3)^{1/2}\exp\{-\kappa\rho(t - 1/\rho)^2/(2t)\}$$

having mean $\mu = 1/\rho$ and coefficient of variation $1/\kappa^{1/2}$. We return to this distribution in Chapters Six and Seven.

2.1.11 The central limit theorem

If $X_1, X_2, X_3 \ldots X_N$ is a sequence of independent, identically distributed random variables with finite means μ and finite non-zero variances σ^2, then the sum $U_N = X_1 + X_2 + X_3 + \ldots + X_N$ has a distribution tending, as $N \to \infty$, to the normal with mean $N\mu$ and variance $N\sigma^2$.

Note that there is no requirement that the random variables $X_i (i = 1, 2, \ldots N)$ be continuous. Provided the X_i are independent with common distribution, whatever it may be (there are some pathological exceptions which are of mathematical interest rather than of practical importance), their sum U_N will have a distribution that is approximately normal, and the larger the value of N, the better the approximation. A further consequence of the central limit theorem is that the mean $V_N = U_N/N$ is distributed approximately as $N(\mu, \sigma^2/N)$, whatever the distribution of the X_i.

2.2 PROBABILITY GENERATING FUNCTIONS

In this section, we consider the discrete random variables X of section 2.1.7 (equation 2.9) such that $P[X = x] = p_x$. We introduce a dummy variable z, and consider the sum

$$P(z) = p_0 + p_1 z + p_2 z^2 + p_3 z^3 + \ldots$$

termed the **probability generating function.** $P(z)$ will end in a term $p_N{}^{zN}$ where the random variable X can take one of $N + 1$ values, as in the binomial distribution, or will be an infinite series, as for the Poisson distribution. (Note that the letter P, here denoting a series whether finite or infinite, is used here in a different sense from that denoting a probability, as in $P[X = x]$.) Since the probabilities p_i sum to one, we must have $P(1) = 1$. Differentiating $P(z)$ with respect to the dummy variable z, we obtain

$$P'(z) = 1.p_1 + 2.p_2 z + 3.p_3 z^2 + \ldots \quad \text{so that} \quad P'(1) = \Sigma x p_x.$$

Denoting by μ the mean of the distribution $P[X = x]$, we see that $\mu = P'(1)$. If the probability $P[X > x]$ is denoted by q_x, so that the distribution function $P[X \leq x] = 1 - q_x$, we can define a generating function $Q(z)$ for the probabilities q_x of the form

$$Q(z) = q_0 + q_1 z + q_2 z^2 + q_3 z^3 + \dots$$

(but note that the sum of the q_x will not be unity) which is related to $P(z)$ by

$$(1 - z)Q(z) = 1 - P(z)$$

as can be verified by comparing the terms on each side. Differentiating this relation with respect to z gives $\mu = P'(1) = Q(1)$, and a second differentiation can be shown to give the variance σ^2 as $\sigma^2 = P''(1) + P'(1) - \{P'(1)\}^2 = 2Q'(1) + Q(1) - \{Q(1)\}^2$.

Probability generating functions are particularly useful where it is required to obtain the probability distribution of a sum of discrete random variables. Consider two **independently distributed, non-negative** random variables, X and Y, such that $P[X = x] = r_x$ and $P[Y = y] = s_y$; denote their probability generating functions by $R(z)$ and $S(z)$ respectively. The probability of the event "$X = x$ and $Y = y$" is then

$$P[(X = x) \cap (Y = y)] = r_x s_y.$$

Now consider the new random variable $U = X + Y$; the event "$U = u$" can occur in the following $u + 1$ mutually exclusive ways: $X = 0, Y = u; X = 1, Y = u - 1; \dots X = u, Y = 0$. Hence,

$$P[U = u] = P[(x = 0) \cap (Y = u)] + P[(X = 1) \cap (Y = u - 1) \\ + \dots P[(X = u) \cap (Y = 0)] = r_0 s_u + r_1 s_{u-1} + \dots + r_u s_0.$$

But this is the cofficient of z^u in the expansion of the product $U(z) = R(z)S(z)$. Therefore, given the probability generating functions of the probabilities $\{r_x\}, \{s_y\}$, the probability generating function of $U = X + Y$ is determined. More generally, given the independently distributed and non-negative discrete random variables $X_1, X_2, \dots X_N$ with probability generating functions $P_1(z), P_2(z), \dots P_N(z)$, the probability generating function of the sum $U = X_1 + X_2 + \dots + X_N$ is $P_1(z)P_2(z)\dots P_N(z)$, and the coefficient of z^u in the expansion of this product gives the probability $P[U = u]$. If the X_i all have the same probability generating function $P(z)$, then the probability $P[U = u]$ is the coefficient of z^u in the expansion of $\{P(z)\}^N$.

As a simple illustration of the above, consider the case of the Bernoulli distribution with $P[X = 0] = q, P[X = 1] = p$, with $p + q = 1$. The probability generating function is $P(z) = q + pz$, and the probability generating function of the sum of N independent Bernoulli trials is $(q + pz)^N$. Expanding in powers of z, we find that the sum U of N independent Bernoulli trials $U_N = X_1 + X_2 + \dots + X_N$ has

$$P[U = u] = N!/[u!(N - u)!]p^u q^{N-u}$$

in accordance with section 2.1.10.

2.2.1 Compound distributions

Probability generating functions have further application where the number N of variables in the sum $U = X_1 + X_2 + \ldots + X_N$ is itself a random variable. Suppose, as before, that the probability generating function of each of the independent X_i is $P(z)$, and suppose that $P[N = n] = v_n$, with probability generating function

$$V(z) = v_0 + v_1 z + v_2 z^2 + v_3 z^3 + \ldots.$$

Using the theorem of total probability, we have

$$P[U_n = l] = \sum_n p[U_n = l \mid N = n].P[N = n]$$

$$= \sum_n v_n.P[U_n = l \mid N = n].$$

But for N fixed at value n,

$$\sum_l P[U_n = l \mid N = n]z^l = \{P(z)\}^l.$$

Hence the probability generating function of U,

$$U(z) = u_0 + u_1 z + u_2 z^2 + u_3 z^3 + \ldots$$

is

$$U(z) = \sum_l u_l z^l = \sum_l z^l \sum_n v_n.P[U_n = l \mid N = n]$$

$$= \sum_n v_n \sum_l P[U_n = l \mid N = n]z^l = \sum_n v_n \{P(z)\}^n = V(P(z)).$$

As an example, consider the distribution of the sum

$$U_N = X_1 + X_2 + X_3 \ldots + X_N$$

of N independent Bernoulli variables, each with

$$P[X = 0] = q, P[X = 1] = p,$$

where N is a random variable having a Poisson distribution with parameter λ, and hence probability generating function $\exp(-\lambda(1 - z))$, as can be verified. For fixed $N = n$, it was shown in the last section that the probability generating function of U_N is $(q + pz)^N$; hence the probability generating function of

$$U_N = X_1 + X_2 + X_3 + \ldots + X_N,$$

where N has a Poisson distribution, is $\exp(-\lambda(1-(q+pz)))$. Hence, by calculating the coefficient of z^u when this expression is expanded in powers of z, we have

$$P[U = u] = \exp(-\lambda p).(\lambda p)^u/u!$$

so that the sum has a Poisson distribution with parameter λp.

This section has given only the briefest of accounts of some of the more important properties of probability generating functions; the reader is referred to Chapter Five of the book by Grimmett and Stirzaker (1992) for further reading.

2.3 GENERALISED LINEAR MODELS

Hitherto, the material of this chapter has not dealt with **data**, although it has been mentioned, for example, that a Poisson process might be adopted as an initial hypothesis for the distribution in time of el Niño occurrences, and that a statistical analysis of data would follow to determine whether the weight of evidence supported this initial hypothesis, or whether it should be rejected in favour of some alternative. The remainder of this chapter deals with some aspects of **statistical estimation** and **hypothesis testing** which will be required in later chapters. The presentation here given affords no more than an introduction; for greater depth of treatment, the reader is referred to McCullagh and Nelder (1989) for a general account. Hosmer and Lemeshow (1989) give a very readable account of generalised linear models using Bernoulli and binomial distributions, and there is much useful information in the manuals for certain statistical packages such as GLIM 4 (1993) and Genstat 5 Release 3 (1993).

An important component of any basic statistical course is linear regression analysis in which a simple regression is denoted by $y_i = \beta_0 + \beta_1 x_i + \varepsilon_i$ or, since y_i is an observed value of the random variable Y_i, by $E[Y_i] = \beta_0 + \beta_1 x_i$. Such courses give the theory by which the parameters β_0 and β_1 are estimated using least squares given N pairs of measurements (x_i, y_i) on the independent or **explanatory variable** x_i, and the dependent or **response variable** $y_i (i = 1 \ldots N)$. On the assumption that the **errors** or **residuals** ε_i have a distribution $N(0, \sigma^2)$, theory is also given in such courses by which F- and t-tests can be used to test hypotheses such as $H_0: \beta_1 = 0$ against the alternative hypothesis $H_1: \beta_1 \neq 0$. Similar tests are also expounded for hypotheses concerning the intercept parameter β_0. Central to such tests is the assumption that the residuals, and hence the observations $y_1, y_2, y_3, \ldots y_N$ of the random variable Y, are independently and normally distributed with constant variance (although some of these assumptions are subsequently relaxed). The elementary statistical courses also discuss **multiple regression**, conveniently represented in matrix notation by $\boldsymbol{y} = \boldsymbol{X\beta} + \boldsymbol{\varepsilon}$, where $\boldsymbol{y}$ is an $N \times 1$ vector of observations of the random variable Y, $\mathbf{X}$ is a $N \times (p+1)$ matrix of observations of the explanatory variables $x_1, x_2, x_3 \ldots x_p$, $\boldsymbol{\beta}$ is a $(p+1) \times 1$ vector of regression coefficients, and $\boldsymbol{\varepsilon}$ is a $N \times 1$ vector of residuals, independently and normally distributed $N(0, \sigma^2)$.

It was stated above that events distributed randomly in time or space follow a Poisson process, in which the number X of events occurring in an interval of length T has probability $P[X = x] = \exp(-\lambda T).(\lambda T)^x/x!$, and that the intervals between events are exponentially distributed. For observations of such variables, the normal assumption is clearly inappropriate. There are also other circumstances in which the multiple regression model gives difficulties; for example, in Chapter Three we wish to model the time dependence of the proportion r/n of days that rain falls on day t, given that day $t-1$ was dry. Here, the response variable Y is a proportion restricted to the interval $(0, 1)$, and a multiple regression model of Y on harmonic functions of time (say, $cos(2\pi t/365), sin(2\pi t/365)$), may result in fitted values $\hat{Y}$ lying outside the permitted range. The same difficulty may arise where a Poisson response variable is regressed upon harmonic functions of time.

The purpose of this section, therefore, is to extend the basic material of multiple regression analysis to cases – like those just mentioned – where the response variable Y has distributions other than the normal, such that fitted values of the response variable must satisfy certain conditions.

2.3.1 The structure of statistical models

As before, a set of observations of a random variable Y is denoted by $y_1, y_2, y_3, \ldots y_N$. Each observation y_i can be regarded as composed of a **systematic component** μ_i, together with a **random error component** ε_i, such that $y_i = \mu_i + \varepsilon_i$. Statistical modelling can be regarded as a process in which the data values y_i are to be matched by the μ_i, and for a "good" model, the μ_i must satisfy two requirements: (1) they must all be expressed in terms of a small number of basic quantities, the **model parameters**; (2) the resulting μ_i must all be close to the data values y_i. Information contained in the data y_i will thereby be brought into sharper focus, since we can concentrate attention on the estimated μ_i, free from the "blur" associated with the random errors ε_i. Modelling requires two decisions: (1) how to define the relation between the μ_i and underlying model parameters; and (2) how to choose a measure of the discrepancy showing how well a given set of μ_i agree with the N data values y_i. The first decision (1) concerns the systematic component of the model, whilst the second decision (2) requires a statistical decision about the part of the variation that the systematic component does not account for.

The generalised linear models (GLMs) used in the following chapters assume a particular model structure.

Concerning the **systematic component**, it is first assumed that there is a **linear predictor**, denoted by η, of the form

$$\eta_i = \sum_j \beta_i x_{ij} \qquad i = 1, 2, \dots N$$

where the x_j are the explanatory variables. In several cases that follow, where we are concerned with representing seasonal variation, the explanatory variables will be periodic functions of time: for example

$$x_{i1} = \cos(2\pi i/T) \qquad x_{i2} = \sin(2\pi i/T)$$
$$x_{i3} = \cos(4\pi i/T) \qquad x_{i4} = \cos(4\pi i/T)$$

for some T, typically $T = 365$ when the y_i are observed at daily intervals.

Second, it is assumed that there is a **link function, of known form**, denoted by $\eta_i = g(\mu_i)$.

A simple example of a known link function is $g(\mu_i) = \mu_i$; another link function used in later chapters is

$$g(\mu_i) = \log_e(\mu_i/(1 - \mu_i)).$$

Thus the two basic assumptions about the systematic component are that η is linear in the unknown parameters β_j, and that $g(.)$ is of known form, but it is fairly straightforward to generalise to cases where the linear predictor η contains parameters that occur non-linearly (for example, where one or more explanatory variables have the form x_j^λ, where the parameters λ are to be estimated); and to the case where $g(.)$ contains unknown parameters (for example, $g(\mu_i) = (\mu_i^\lambda - 1)/\mu_i^\lambda$, again with λ to be estimated). Extensions of this form are not used in this book.

Regarding the **random error component**, each observation y_i of the random variable Y is assumed to have a pdf belonging to the **exponential family** of distributions (not to be confused with the exponential distribution itself, which also is a member of the exponential family). The importance of the exponential family is that for every distribution belonging to it, **sufficient statistics** can be calculated which utilise all the information contained in the data about the distribution's parameters; having calculated these sufficient statistics, the raw data could in effect be thrown away (provided, of course, that the analyst is sufficiently confident that the raw data do really come from the distribution that he has specified!). The exponential family of distributions all have pdfs that can be expressed in the following general form:

$$f(y_i; \theta_i, \phi) = \exp\{(y_i\theta_i - b(\theta_i))/a_i(\phi) + c(y_i, \phi)\}.$$

The functions $a_i(\phi)$, $b(\theta_i)$ and $c(y_i, \phi)$ determine the particular member of the family of distributions from which y_i is a sample. Although the form of the exponential family appears unusual at first sight, it can be shown that it includes the Bernoulli, binomial, Poisson, gamma, inverse gamma and normal distributions as special cases. Note that the parameter θ can vary with i, and depends upon the the parameters β_j through the link function; however, the parameter ϕ, termed the **scale parameter**, does not vary with i. In fact the mean $E[Y_i] = \mu_i = b'(\theta_i)$ depends only on the θ parameter, whilst the variance $\text{var}[Y_i] = a_i(\phi)b''(\theta_i)$ can be written in the form $\phi.V(\mu_i)$, where $V(.)$ is the **variance function**, depending only on μ_i.

To see how, say, the binomial distribution is a member of the exponential family, we can write it in the form

$$\begin{aligned} f(y) &= N!/[y!(N-y)!]\pi^y(1-\pi)^{N-y} \\ &= \exp\{y.\log_e[\pi/(1-\pi)] + N.\log_e(1-\pi) + \log_e(N!/[y!(N-y)!])\} \end{aligned}$$

so that the θ parameter in this case is $\log_e[\pi/(1-\pi)]$,

$$\begin{aligned} a_i(\phi) &= 1, \\ b(\theta) &= -N\log_e(1-\pi) = N\log_e[1+\exp(\theta)] \text{ and} \\ & \qquad c(y,\phi) = \log_e(N!/[y!(N-y)!]). \end{aligned}$$

Equally, the Poisson distribution may be written

$$f(y) = \exp\{y\log_e\lambda - \lambda - \log_e(y!)\}$$

so that the θ parameter is $\log_e\lambda$,

$$a_i(\phi) = 1, \quad b(\theta) = \exp(\theta) \quad \text{and} \quad c(y, y, \phi) = -\log_e(y!).$$

In the case of the binomial and Poisson distributions, we have seen that the scale parameter ϕ can be given the value 1; this is because these two distributions are determined by a single parameter (π in the case of the binomial, λ in the case of the Poisson). The normal distribution on the other hand, written as

$$f(y) = \exp\{(y\mu - \tfrac{1}{2}\mu^2)/\sigma^2 - \tfrac{1}{2}(\log_e(2\pi\sigma^2) + y^2/\sigma^2)\}$$

has the θ parameter equal to μ, and the scale parameter $\phi = \sigma^2$. A fortunate characteristic of GLMs is that the parameters β_j, and hence the means μ_i, can be estimated first without having to estimate the scale parameter at the same time. However, we are usually interested not only in estimating the μ_i but also in testing hypotheses about the model structure, and for this purpose an estimate of the scale parameter ϕ is essential. In practice, an estimate of ϕ is given by the deviance statistic D – defined below – and its corresponding degrees of freedom d: $\hat{\phi} = D/d$.

Thus we have seen that, by introducing (1) the link function $\eta_i = g(\mu_i)$, where η_i is the linear predictor $\eta_i = x_i^T\beta$, and (2) the exponential family of distributions for the observed values $y_1, y_2, y_3 \ldots y_N$, we have greatly extended the range of possible models used to describe the characteristics of data, the familiar multiple regression model being a special case of the more general class of models.

2.3.2 The likelihood function, and its role in estimation GLM parameters

Suppose we have a pdf written as $f(y_i, \mu_i, \phi)$, the distribution of a discrete random variable Y_i being determined by the two parameters μ_i and ϕ. When the observations y_i are sampled randomly from their pdfs, the events $Y = y_1$, $Y = y_2, \ldots$ are independent, so that the probability of obtaining the observed set is

$$f(y_1, \mu_1, \phi).f(y_2, \mu_2, \phi)\ldots.f(y_N, \mu_N, \phi).$$

For continuous random variables, the corresponding probability is

$$f(y_1, \mu_1, \phi).f(y_2, \mu_2, \phi)\ldots.f(y_N, \mu_N, \phi)\mathrm{d}y_1\mathrm{d}y_2\ldots\mathrm{d}y_N.$$

The **likelihood function** is defined as a function of the unknown parameters to be estimated as follows:

$$L(\mu_1, \mu_2, \mu_3 \ldots \mu_N, \phi; y_1 y_2, \ldots y_N) = \prod_{i=1}^{N} f(y_i, \mu_i, \phi).$$

Maximum likelihood estimates of the parameters $\mu_1, \mu_2, \ldots \mu_N, \phi$ are those values $\hat{\mu}_1, \hat{\mu}_2, \hat{\mu}_3 \ldots \hat{\mu}_N, \hat{\phi}$ which satisfy the condition

$$L(\hat{\mu}_1, \hat{\mu}_2, \ldots \hat{\mu}_N, \hat{\phi}) \geq L(\mu_1, \mu_2, \ldots \mu_N, \phi)$$

and which therefore maximise the probability of the observed data values $y_1, y_2, y_3, \ldots y_N$. For the statistical models in this book, all of which have pdfs belonging to the exponential family, the likelihood function has a unique maximum and no minima, so that the maximum likelihood estimates of model parameters may be found by solving the set of equations obtained by setting the partial derivatives of L (or more conveniently $\log_e L$) to zero:

$$\partial \log_e L/\partial\mu_1 = 0; \qquad \partial \log_e L/\partial\mu_2 = 0; \qquad \ldots \partial \log_e L/\partial\mu_N = 0;$$
$$\partial \log_e L/\partial\phi = 0.$$

The matrix of second derivatives $I(\hat{\theta}) = -\partial^2 \log_e L/\partial\theta_i\partial\theta_j$ (where θ_j, θ_j are pairs from the set $\mu_1, \mu_2, \ldots \mu_N, \phi$) evaluated at the maximum likelihood estimate $\hat{\mu}_1, \hat{\mu}_2, \ldots \hat{\mu}_N, \hat{\phi}$ plays a central role in assessing the accuracy of the maximum likelihood estimates.

2.3.3 Goodness of model fit: the scaled deviance

Given the data $y_1, y_2, y_3, \ldots y_N$, we commonly need to know how many explanatory variables x_j need to be included in the linear predictor

$$\eta = x^T\beta = \sum_{j=1}^{p} \beta_j x_j.$$

For example, where the explanatory variables are harmonic functions of time, perhaps $x_1 = \cos(2\pi t/T), x_2 = \sin(2\pi t/T)\ldots$, we may wish to know whether it is necessary to include higher order terms, $x_3 = \cos(4\pi t/T)$ and $x_4 = \sin(4\pi t/T)$. At one extreme, we could include $p = N$ explanatory variables (always supposing that N linearly independent explanatory variables could be found) and since we should then be fitting N parameters β_j to N data values, the maximum likelihood estimates of the μ_i are then simply the observations themselves, $\hat{\mu}_i = y_i$. This is known as the **full model**, for which the maximised value of the log-likelihood function is denoted by $\log_e L_f$. At the other extreme, we could include no explanatory variables x_j, which is equivalent to putting $\mu_1 = \mu_2 = \ldots = \mu_N = \mu$, a constant. This other extreme is termed the **null model**; and obviously in all but trivial cases the null model will fail to give an adequate description of the data. Intermediate between these extremes are models which are of more practical interest: the largest, most complex model (in terms of the number of explanatory variables to be included) is the **maximal model**, whilst a corresponding **minimal model** can also be defined (containing the very smallest number of explanatory variables, and hence parameters β_j, to be included). The model between the maximal and minimal models that is currently being considered for its adequacy as a description of the data, is termed the **current model**, for which the maximised value of the log-likelihood function is denoted by $\log_e L_c$.

An intuitively obvious method of comparing the fit of two models, say the current model relative to the full model, is in terms of the ratio of their likelihoods, remembering that the likelihood is proportional to the probability of the sample $y_1, y_2, y_3 \ldots y_N$ given the model structure; if the current model is a poor description of these data L_c will be small relative to L_f, so that the ratio L_c/L_f will be close to zero. Therefore, $-\log_e(L_c/L_f)$ will be a large positive number. Conversely, if the current model provides almost as good a fit as the full model, L_c will be only slightly less than L_f, so that the ratio L_c/L_f will be only slightly less than one, and $-\log_e(L_c/L_f)$ will be small and positive. The statistic

$$S(c,f) = -2\log_e(L_c/L_f)$$

is termed the **scaled deviance**; when expressed in the form

$$S(c,f) = D(c,f)/\phi$$

then the numerator $D(c,f)$ is the **deviance statistic** (so that the scaled deviance is equal to the deviance $D(c,f)$ divided, or scaled, by the scale parameter ϕ). The deviance is a measure of lack of model fit; the larger the deviance, the poorer is the current model as a description of the data $y_i (i = 1, 2, 3, \ldots N)$. The advantage of using the deviance, rather than the scaled deviance, is that it does not contain the unknown scale parameter ϕ, and can therefore be computed from the data. In the particular case where each y_i has a normal distribution and the link function is the identity link $g(\mu) = \mu$, the deviance $D(c,f)$ reduces to the residual sum of squares, and $S(c,f)$ becomes the ratio

$$[\text{residual sum of squares}]/\sigma^2$$

which is shown by statistical theory to follow a χ^2 distribution with the degrees of freedom associated with the numerator. It is straightforward to find expressions for the deviance for other distributions in the exponential family: for the Poisson, for example,

$$\log_e f(y_i) = y_i \log_e \lambda_i - \lambda_i - \log_e(y_i!)$$

so that

$$\log_e L = \Sigma y_i \log_e \lambda_i - \Sigma \lambda_i - \Sigma \log_e(y_i!).$$

But in the case of the full model, the estimate of each λ_i is y_i, so that the log likelihood for the full model, namely $\log_e L_f$, becomes

$$\log_e L_f = \Sigma y_i \log_e y_i - \Sigma y_i - \Sigma \log_e(y_i!).$$

Also, for a current model with log-likelihood $\log_e L_c$ we have

$$\log_e L_c = \Sigma y_i \log_e \hat{\lambda}_i - \Sigma \hat{\lambda}_i - \Sigma \log_e(y_i!)$$

and the deviance $-2\log_e(L_c/L_f)$ has the form

$$2\Sigma\{y_i \log_e(y_i/\hat{\lambda}_i) - (y_i - \hat{\lambda}_i)\},$$

the form used in the example given in section 2.3.4 below. Similar calculations give explicit formulae for the deviances of the binomial, gamma and inverse gamma distributions.

Statistical theory also provides information about the distribution of S in the case where the y_i have distributions other than the normal, but belonging to the exponential family. If one model (say Model 1) is "nested" within another (Model 2, say) in the sense that for Model 2 the linear predictor is

$$\eta^{(2)} = \beta_1 x_1 + \beta_2 x_2 + \beta_3 x_3 + \ldots + \beta_{p+q} x_{p+q} \quad (q > 0)$$

whilst that for Model 1 is

$$\eta^{(1)} = \beta_1 x_1 + \beta_2 x_2 + \beta_3 x_3 + \ldots + \beta_p x_p$$

then

$$S = -2\log_e\{\text{maximised likelihood for Model 1/maximised likelihood for Model 2}\}$$

has a distribution that is approximately χ^2 with q degrees of freedom, the difference between the number of parameters fitted in Model 2 and the number of parameters fitted in Model 1. There is a practical difficulty in using such a test in that S requires knowledge of the unknown scale parameter ϕ. To circumvent this difficulty, it is common to fit a maximal model, as defined above, with say $p + q + r$ parameters β_i; then if D (Model 2, maximal model) is the deviance statistic when Model 2 is fitted instead of the maximal model, an estimate of ϕ is D (Model 2, maximal model)/r. Alternatively, we can avoid the problem by calculating ratios of scaled deviances, since the scale parameter will then disappear; thus if

$$S\,(\text{Model 2, maximal model}) = D\,(\text{Model 2, maximal model})/\phi$$

is the scaled deviance when Model 2 is fitted and

$$S\,(\text{Model 1, Model 2}) = D\,(\text{Model 1, Model 2})/\phi$$

when the more restricted Model 1 is subsequently fitted, then the ratio

$$S\,(\text{Model 1, Model 2})/S\,(\text{Model 2, maximal model})$$

does not involve ϕ. If Model 1, Model 2 and the maximal model contain p, $p + q$, and $p + q + r$ parameters, then

$$D\,(\text{Model 1, Model 2})/q$$

divided by

$$D\,(\text{Model 2, maximal model})/r$$

is taken to have an approximate F distribution with q and r degrees of freedom.

Thus the deviance statistic is a measure of the goodness of model fit, and ratios of deviances provide approximate F-tests by which we can test whether a model with fewer parameters (fewer β_i) can be used instead of a more complex model (with more β_i). In the following chapter, the deviance is used to test whether a model with y_i binomially distributed and link function $g(u) = \log_e[\mu/(1 - \mu)]$ is satisfactory when two explanatory variables

$$x_{i1} = \cos(2\pi i/T) \quad \text{and} \quad x_{i2} = \sin(2\pi i/T)$$

adequately represent the seasonal variation in the y_i, or whether additional terms such as

$$x_{i3} = \cos(4\pi i/T) \quad \text{and} \quad x_{i4} = \cos(4\pi i/T)$$

also need to be included in the model.

2.3.4 Example

As a small example, consider the reconstructed sequence of el Niño events. Neglecting the important fact that the sequence is incomplete for the sixteenth and twentieth centuries, the numbers of events in each century were 2, 3, 7, 4 and 2; and we suppose that an explanatory variable (artificially constructed, for the purpose of illustration) has values {28, 39, 29, 70, and 7}. A reasonable starting-point would be to assume that the number of events per century has a Poisson distribution, and we might wish to explore whether its mean value depends upon the values of the explanatory variable, which we call x. Denoting by Y the Poisson random variable with observed values

$$y_i\{y_1 = 2, y_2 = 3, y_3 = 7, y_4 = 4, y_5 = 2\}$$

we might wish to consider the nested models

$$\mathrm{E}[Y_i] = \lambda_i \qquad \eta_i^{(2)} = \beta_0 \qquad \eta_i^{(1)} = \beta_0 + \beta_1 x_i$$

with link function $g(\lambda_i) = \log_e \lambda_1$. The case $\eta_i^{(2)} = \beta_0$ is the null model, with $\lambda_1 = \lambda_2 = \ldots = \lambda$, a constant. For the Poisson distribution, the scale parameter can be taken to be one, and the scaled deviance, equal to the deviance, has value 4.356 on 4 degrees of freedom. The estimate of β_0 is 1.281 ± 0.236, so that the estimate of the (constant) mean λ is $\hat{\lambda} = \exp(1.281) = 3.6002$.

This null model, however, is simply equivalent to fitting a Poisson distribution with parameter λ to the data y_i by maximum likelihood, for which the well-known estimate is

$$\hat{\lambda} = (2 + 3 + 7 + 4 + 2)/5 = 3.6.$$

Of greater interest is the model

$$\mathrm{E}[Y_i] = \lambda_i \qquad g(\lambda_i) = \log_e \lambda_i = \eta_i^{(1)} \qquad (i = 1 \ldots 5) \qquad \eta_i^{(1)} = \beta_0 + \beta_1 x_i$$

where the Poisson mean λ_i is taken to depend on the explanatory variable x_i. Fitting this model, it is found that the inclusion of the additional parameter β_1 decreases the deviance (= scaled deviance) by 0.287 to a value of 4.069. The analysis may be expressed in terms of the following table (analysis of deviance table):

	df	Deviance	Mean deviance	Deviance ratio
Regression on x	1	0.287	0.287	0.29
Residual	3	4.069	1.356	
Total	4	4.356	1.089	
Change	− 1	− 0.287	0.287	0.29

Estimates of the parameters β_0 and β_1 are now as follows:

Parameter	Estimate	Estimate/S.E.
β_0	1.065 ± 0.477	2.23
β_1	0.0060 ± 0.0111	0.54

from which it is clear that the estimate of β_1 is small in relation to its standard error, and we have no evidence that the mean of the Poisson distribution is related to the variable x.

Instead of working with the number of el Niño occurrences per century, we might prefer to work with the intervals between events in the reconstructed series. These intervals have values denoted by $y_i (i = 1 \ldots 17)$ with the 17 values

$$\{37, 36, 10, 28, 49, 19, 8, 19, 16, 7, 21, 13, 10, 14, 63, 34, 57\};$$

assuming that the y_i come from gamma distributions with means μ_i, the model being

$$E[Y_i] = \mu_i \qquad g(\mu_i) = 1/\mu_i = \eta_i^{(1)} \qquad (i = 1 \dots 17) \qquad \eta_i^{(1)} = \beta_0 + \beta_1 x_i$$

where the candidate explanatory variable x (artificially constructed) has the values

$$\{26, 62, 38, 97, 75, 84, 16, 7, 44, 99, 83, 11, 46, 32, 24, 20, 14\}.$$

When the null model is fitted ($\beta_1 = 0$, corresponding to fitting a gamma distribution by maximum likelihood to the 17 y_i values), the deviance is found to be 7.355 on 16 degrees of freedom, so that the estimate of the scale parameter ϕ is

$$\hat{\phi} = 7.355/16 = 0.460.$$

The estimate of β_0 is

$$\hat{\beta}_0 = 0.03855 \pm 0.00634,$$

so that the common mean $\mu_i = \mu$ is estimated from the link function as

$$\hat{\mu} = 1/0.03855 = 25.9403.$$

The null model simply fits a gamma distribution to the y_i, and the maximum likelihood estimate of the gamma mean is simply the mean of the y_i, namely $(37 + 36 + 10 + \dots + 57)/17 = 25.9412$, almost identical to the value obtained using the null model.

The null model is of little interest, however, and we wish to determine whether the gamma mean μ_i depends upon the candidate explanatory variable x. Instead of the null model $\eta_i^{(2)} = \beta_0$, we therefore use instead the linear predictor $\eta_i^{(1)} = \beta_0 + \beta_1 x_i$; fitting the additional parameter β_1 reduces the deviance by 0.135, and the estimates of the two parameters β_0, β_1 are given by

Parameter	Estimate
β_0	0.0336 ± 0.0160
β_1	0.000116 ± 0.000321

Clearly the estimate of β_1 is small relative to its standard error, giving no evidence that the gamma mean μ_i is related (through the link function) to the candidate variable x.

2.4 CONCLUSION

It will be clear that GLMs are an essential tool for exploring the nature of the relationships between model parameters and variables which might explain whether and how the parameters change through time and space. The following chapter introduces the reader to an important application of GLMs in which daily rainfall records are analysed to answer questions of importance in agricultural planning, particularly in sem-arid areas where the correct determination of planting date is of critical importance.

Chapter Three

Markov chains: applications to modelling of daily rainfall

3.1 SIMPLE BERNOULLI SEQUENCES AND THEIR GENERALISATION TO MARKOV CHAINS

If we throw a coin repeatedly, with the outcome of each toss independent of the preceding tosses, we obtain a sequence of 0s and 1s, where 0 represents the outcome "coin shows heads", and 1 represents the outcome "coin shows tails" (we ignore the possibility of other outcomes, such as the coin landing on its edge, or disappearing). If the coin is unbiased, the proportion of 0s in the sequence will tend, in probability, to the value 0.5. Similarly, if we choose arbitrarily a sequence of N values $x_1, x_2, x_3 \ldots$ from a table of random numbers and record as a "success" a value of x_i greater than, say, 0.8, we obtain a sequence of observations of a random variable Y having just two outcomes ($Y = 1$ for a "success", $Y = 0$ for a "failure"). Such sequences, in which the occurrence of the outcome of the ith event is independent of all previous outcomes, are simple **independent Bernoulli sequences**. For any two consecutive events, say $Y_{i-1} = 1$ and $Y_i = 0$, their independence may be expressed formally by

$$P[(Y_{i-1} = 1) \cap (Y_i = 0)] = P[Y_{i-1} = 1].P[Y_i = 0] \qquad (3.1)$$

so that the joint probability of the two events is simply the product of the two events taken separately, by the definition of independence given in Chapter Two. Generalising (3.1), we have

$$P[(Y_i = k_i) \cap (Y_{i-1} = k_{i-1}) \cap (Y_{i-2} = k_{i-2}) \ldots \cap (Y_0 = k_0)] = \prod_{j=1}^{i} P[Y_j = k_j] \qquad (3.2)$$

where each of the k_j ($j = 0, 1, 2 \ldots i$) takes values 0 or 1, and the product on the right-hand side is over values of $j = 0, 1, \ldots i$.

The physical world also yields sequences of observations of a random variable Y taking values 0 and 1. Examples are:

1. The occurrence ($Y = 1$) or non-occurrence ($Y = 0$) of rain on a particular day;
2. The occurrence ($Y = 1$) or non-occurrence of a monthly flow in a river greater than the median flow fow that month;
3. The occurrence ($Y = 1$) or non-occurrence of surface flow at a particular point on a drainage basin, at any instant in time.

In examples (1) and (2), a sequence of days (months) yields a sequence consisting of 0s or 1s ; in (3), a set of simultaneous measurements made at points with different spatial coordinates **s** in the drainage basin yields a two-dimensional array of data taking values 0 or 1. In many such examples from the physical world, however, the relation (3.1) will not hold; if it rained yesterday because of the slow passage of a frontal system, clearly it is more likely to rain today also; and if yesterday was dry because of highly stable meteorological conditions, it is very likely that there will be no rain today either. The sequence of 0s and 1s is then a **dependent Bernoulli sequence**; in place of (3.1), therefore, we have a relation involving conditional probabilities

$$P[(Y_i = 0) \cap (Y_{i-1} = 1)] = P[Y_i = 0 \mid Y_{i-1} = 1].P[Y_{i-1} = 1] \quad (3.3)$$

and, in place of (3.2), we have

$$\begin{aligned} P[(Y_i = k_i) \cap (Y_{i-1} = k_{i-1}) \cap (Y_{i-2} = k_{i-2}) \ldots \cap (Y_0 = k_0)] \\ = P[(Y_i = k_i) \mid (Y_{i-1} = k_{i-1}) \cap (Y_{i-2} = k_{i-2}) \ldots \cap (Y_0 = k_0)]. \\ P[(Y_{i-1} = k_{i-1}) \mid (Y_{i-2} = k_{i-2}) \ldots \cap (Y_0 = k_0)] \ldots \\ P[(Y_1 = k_1) \mid (Y_0 = k_0].P[Y_0 = k_0]. \end{aligned} \quad (3.4)$$

Suppose, however, that all the past history of the sequence $Y_{i-1}, Y_{i-2}, \ldots Y_0$ can be regarded as summarised by the value taken by the random variable Y at time $(i - 1)$: then (3.4) simplifies to

$$\begin{aligned} P[(Y_i = k_i) \cap (Y_{i-1} = k_{i-1}) \cap (Y_{i-2} = k_{i-2}) \ldots \cap (Y_0 = k_0)] \\ = P[(Y_i = k_i) \mid (Y_{i-1} = k_{i-1})].P[(Y_{i-1} = k_{i-1}) \mid (Y_{i-2} = k_{i-2})] \ldots \\ P[(Y_1 = k_1) \mid (Y_0 = k_0)].P[Y_0 = k_0]. \end{aligned} \quad (3.5)$$

When such simplification is possible, the sequence $Y_j (j = 1, 2, \ldots i)$ is said to be a **first-order Markov process**, the essential point of which is that knowledge of the system at time $i - 1$ summarises the history of the process up to that time; or expressed in other words, knowledge of the system at time i, conditional on knowledge of how the system stood at times $i - 1, i - 2 \ldots$ is simply equal to knowledge of the system at time i, conditional only on knowledge of how it stood at time $i - 1$. Since days are either dry or wet, the values 0 and 1 taken by the random variable Y_j are the **states** of the process; the random variable Y_j makes **transitions** between the states, and these transitions have conditional probabilities $P[(Y_i = k_i)|(Y_{i-1} = k_{i-1})]$ which can be arranged in a **matrix of transition probabilities**, of the form

$$\mathbf{P} = \begin{bmatrix} P[(Y_i = 0)|(Y_{i-1} = 0)] & P[(Y_i = 1)|(Y_{i-1} = 0)] \\ P[(Y_i = 0)|(Y_{i-1} = 1)] & P[(Y_i = 1)|(Y_{i-1} = 1)] \end{bmatrix}. \qquad (3.6)$$

Assuming that the conditional probabilities in (3.6) do not change in time (in technical terms, a Markov chain is said to be **homogeneous** if

$$P[(Y_i = k)|(Y_{i-1} = m)] = P[(Y_1 = k)|(Y_0 = m)] \quad \text{for all } i)$$

the matrix **P** of transition probabilities can be written in terms of constants p_{00}, p_{01}, p_{10} and p_{11} as follows:

$$\mathbf{P} = \text{State at time } i-1 \begin{matrix} \\ 0 \\ 1 \end{matrix} \overset{\text{State at time } i}{\overset{\begin{matrix} 0 & \quad 1 \end{matrix}}{\begin{bmatrix} p_{00} & p_{01} \\ p_{10} & p_{11} \end{bmatrix}}}.$$

Since a transition from state 0 (or equally from state 1) at time $i-1$ must be either to state 0 or to state 1 at time i, the transition probabilities p_{00}, p_{01}, p_{10} and p_{11} must satisfy $p_{00} + p_{01} = 1, p_{10} + p_{11} = 1$. Furthermore, the values $p_{01} = 0, p_{10} = 0$ are excluded, since these values would result in a sequence solely of 0s or 1s, depending on the initial state; and the values $p_{01} = 1, p_{10} = 1$ are also excluded, since these values would result in sequences 01010101... or 10101010..., depending on initial state.

In some applications, the sequence of 0s and 1s from a two-state Markov process may be of interest in its own right; for example, in terms of the occurrence or non-occurrence of daily rainfall, the data sequence may be used as shown below to calculate probabilities of runs of dry or wet days. In other applications, however, the occurrence of a "success" ($Y = 1$) may be associated with the value of another random variable Z, say. Thus, on a day with rain ($Y = 1$) we would probably also be interested in the depth of rainfall on that day.

3.2 ESTIMATION OF THE TRANSITION PROBABILITIES FROM DATA: $p_{00}, p_{01}, p_{00}, p_{01}$ ASSUMED TIME INVARIANT

Suppose that $p_{00}, p_{01}, p_{00}, p_{01}$ are constant in time, and are to be estimated from an observed sequence 0110001010011. Here we have a sequence of 13 states, with 12 transitions following the initial state of 0. The sequence contains three transitions from state 0 at time $i-1$ to state 0 at time i; four transitions from state 0 to state 1; three transitions from state 1 to state 0; and two transitions from state 1 to state 1, making 12 altogether. Thus in terms of **numbers** of transitions, we have

$$\mathbf{N} = \text{state at time } i-1 \begin{array}{c} \\ 0 \\ 1 \end{array} \overset{\text{State at time } i}{\begin{array}{c} \begin{array}{cc} 0 & 1 \end{array} \\ \begin{bmatrix} 3 & 4 \\ 3 & 2 \end{bmatrix} \end{array}}$$

so that estimates of the four transition probabilities are

$$\text{State at time } i-1 \begin{array}{c} \\ 0 \\ 1 \end{array} \overset{\text{State at time } i}{\begin{array}{c} \begin{array}{cc} 0 & \quad\quad 1 \end{array} \\ \begin{bmatrix} p_{00} = 3/7 & p_{01} = 4/7 \\ p_{10} = 3/5 & p_{11} = 2/5 \end{bmatrix} \end{array}}.$$

A classical example concerns the study of rainfall occurrence during the wet season of December to February inclusive, at Tel Aviv (Gabriel and Neumann, 1957, 1962). Using data from 27 years of record, the 2437 days showed that a dry day ($Y_i = 0$) followed a dry day ($Y_{i-1} = 0$) in 1049 transitions; a wet day ($Y_i = 1$) followed a dry day ($Y_{i-1} = 0$) in 350 transitions; a dry day ($Y_i = 0$) followed a wet day ($Y_i = 1$) in 351 transitions; and a wet day ($Y_i = 1$) followed a wet day ($Y_i = 1$) in 687 transitions. Hence the estimates of the transition probabilities were

$$\mathbf{P} = \text{State at time } i-1 \begin{array}{c} \\ 0 \\ 1 \end{array} \overset{\text{State at time } i}{\begin{array}{c} \begin{array}{cc} 0 & \quad\quad\quad 1 \end{array} \\ \begin{bmatrix} p_{00} = 1049/1399 & p_{01} = 350/1399 \\ p_{10} = 351/1038 & p_{11} = 687/1038 \end{bmatrix} \end{array}}.$$

By limiting their analysis only to wet season conditions at Tel Aviv, Gabriel and Neumann were able to assume that their Markov process was homogeneous. Where sequences of daily rainfall throughout the entire year are to be analysed, this assumption will rarely be valid, and later sections in this chapter discuss the non-homogeneous case.

3.3 SOME THEORETICAL RESULTS FOR A TWO-STATE MARKOV PROCESS WITH TIME-INVARIANT TRANSITION PROBABILITIES $p_{00}, p_{01}, p_{00}, p_{01}$

Characteristics of two-state Markov processes with time-invariant transition probabilities have been widely studied, and here we quote without proof some of the more important results; the reader is referred to Cox and Miller (1965) for further detail.

1. If the row vector $\boldsymbol{p}^{(n)} = [p_0^{(n)}, p_1^{(n)}]$ denotes the probabilities of the system being in states 0, 1 at time n, with initial states given in the row vector $\boldsymbol{p}^{(0)} = [p_0^{(0)}, p_1^{(0)}]$, then

$$\boldsymbol{p}^{(n)} = \boldsymbol{p}^{(n-1)}\mathbf{P} \tag{3.7}$$

where $\boldsymbol{P}$ is the matrix of transition probabilities. Hence, by repeated substitution,

$$\boldsymbol{p}^{(n)} = \boldsymbol{p}^{(n-1)}\mathbf{P} = \boldsymbol{p}^{(n-2)}\mathbf{P}^2 = \ldots = \boldsymbol{p}^{(0)}\mathbf{P}^n \tag{3.8}$$

The matrix $\mathbf{P}^n$ has elements which are the n-**step transition probabilities**

$$p_{jk}^{(n)} = P[\text{state } k \text{ at time } n \,|\, \text{state } j \text{ at time } 0].$$

(Note that $p_{jk}^{(n)}$ is the probability that the system is in state k, **not necessarily for the first time**, at time n, given that it was in state j at time 0; below, we consider the probability $f_{jk}^{(n)}$ that the system is in state k at time n, for the first time, given that it was in state j at time 0.)

2. The matrix $\mathbf{P}^n$ of n-step transition probabilities is given explicitly by

$$\mathbf{P}^n = (p_{01}+p_{10})^{-1}\begin{bmatrix} p_{10} & p_{01} \\ p_{10} & p_{01} \end{bmatrix} + (p_{01}+p_{10})^{-1}(1-p_{10}-p_{01})^n \begin{bmatrix} p_{01} & -p_{01} \\ -p_{10} & p_{10} \end{bmatrix}.$$

3. A general way of calculating the matrix $\mathbf{P}^n$ of n-step transition probabilities is to use a standard result from matrix algebra which gives a decomposition of the matrix $\mathbf{P}$. Assuming that this matrix has distinct eigenvalues, $\mathbf{P}$ can be expressed in the form

$$\mathbf{P} = \mathbf{Q}\boldsymbol{\Lambda}\mathbf{Q}^{-1}$$

where $\boldsymbol{\Lambda}$ is the matrix of eigenvalues (in the case under consideration, which has only two eigenvalues $\lambda_1, \lambda_2, \Lambda = \text{diag}(\lambda_1, \lambda_2)$) and $\mathbf{Q}$ is a matrix having the eigenvectors $\boldsymbol{e}_1, \boldsymbol{e}_2, \ldots$ as its columns. In the more general case with r states, $\Lambda = \text{diag}(\lambda_1, \lambda_2, \ldots \lambda_r)$ and $\mathbf{Q} = [\boldsymbol{e}_1, \boldsymbol{e}_2, \ldots \boldsymbol{e}_r]$.

4. With two states 0 and 1 (dry and wet, say), suppose that we are initially in state 0, a dry day. If the random variable T is the number of days taken for the first return to state 0 after an unbroken run of wet days (system in state 1), denote by $f_{00}^{(n)}$ the probability that $T = n$ days. Then $T - 1$ is the time spent continuously in state 1, the length of the wet spell. Then

$$f_{00}^{(n)} = P[T = n] = p_{01}.p_{11}^{n-2}.p_{10} \quad \text{for } n = 2, 3, 4, \ldots$$

and $f_{00}^{(1)} = p_{00}$. It follows that the mean length of wet spell is

$$\text{E}[T] = \Sigma n f_{00}^{(n)} = (p_{01} + p_{10})/p_{10}.$$

Similarly, given that we are initially in state 1 (a day with rain) we can calculate the probability of a return to state 1 at time n, after an unbroken run of $T - 1$ dry days (in state 0) as

$$f_{11}^{(n)} = p_{10}.p_{00}^{n-2}.p_{01} \qquad n = 2, 3, 4, \ldots$$

with $f_{11}^{(1)} = p_{11}$. Whence the expected length of dry spell is $(p_{10} + p_{01})/p_{01}$. Variances of the lengths of wet and dry spells, as defined above, can be calculated from the expression $\text{var}[T] = \text{E}[(T - \text{E}[T])^2]$.

The assumption of homogeneity (time invariance of the transition probabilities) is basic to the above results. In the case of daily rainfall, the assumption of homogeneity may be appropriate for relatively short intervals of time, but for sites where rainfall is strongly seasonal it would clearly be invalid to assume time invariance in the transition probabilities throughout the entire year. (The classical model of daily rainfall at Tel Aviv by Gabriel and Neumann, cited above, was restricted to the wet season between December and February, when the assumption of time invariance was warranted.) Therefore, it is necessary to adapt the theoretical results given above to deal with non-stationarity in the transition probabilities, and a series of papers by Stern and his colleagues (1982, 1984) demonstrate how these modifications can be achieved, and how the modified models of daily rainfall are used to answer practical questions arising in agricultural planning.

3.4 INCORPORATING SEASONALITY INTO TRANSITION PROBABILITIES FOR THE OCCURRENCE OF RAINFALL

In this section we continue to consider two states 0 and 1, which we call "dry" and "wet", but allow for the time-variance in the transition probabilities. We also introduce a further generalisation; hitherto, we have considered that a "first-order" Markov chain represented satisfactorily the structural dependence between days with or without rain: that is, we assumed (see (3.4) and (3.5)) that

$$\begin{aligned} P[(Y_i = k_i) \mid (Y_{i-1} = k_{i-1}) \cap (Y_{i-2} = k_{i-2}) \ldots \cap (Y_0 = k_0)] \\ = P[(Y_i = k_i) \mid (Y_{i-1} = k_{i-1})], \end{aligned}$$

with all information about the occupancy of states prior to time $i-1$ being encapsulated in the knowledge of the state occupied at that time. In other words, the probability of the system being in a particular state at time i depended only on the state occupied at time $i-1$: hence the term **first-order Markov chain**. In the real world, however, we may have to admit the possibility that the probability of being in state 1 at time i may depend on what happened in the two preceding days: in symbols,

$$\begin{aligned} P[(Y_i = k_i) \mid (Y_{i-1} = k_{i-1}) \cap (Y_{i-2} = k_{i-2}) \ldots \cap (Y_0 = k_0)] \\ = P[(Y_i = k_i) \mid (Y_{i-1} = k_{i-1}) \cap (Y_{i-2} = k_{i-2})], \end{aligned}$$

giving a **second-order Markov chain**. Similarly, generalisations to third-order,... Markov chains are possible; obviously, we use the lowest-order Markov chain that is consistent with the data, and a later section will demonstrate how to assess whether a first-order Markov chain is adequate, or whether the data show that a second-order chain is necessary to model the occurrence of rainfall satisfactorily.

Suppose, for illustration, that we have an N-year record of daily rainfall, and for simplicity let us assume that the record is complete with no days' observations missing (if some of the daily data are missing, this does not complicate matters: it is just that the method becomes more difficult to describe). Let us suppose, also, that we wish to fit a second-order Markov chain to the sequence of 0's and 1's representing the non-occurrence or occurrence of rainfall: that is, we allow the occurrence of state (0 or 1) on, say, 3 January to depend on the states on 2 January and 1 January. More generally, the occurrence of state $j(j = 0$ or $1)$ on day t, is to depend on the occurrence of state $i(i = 0$ or $1)$ on day $t - 1$, and also on the occurrence of state $h(h = 0$ or $1)$ on day $t - 2$. Since the N years of record are complete, there will be N triplets h, i, j with h, i, j taking the values 0 or 1, which describe the sequence of three states up to and including day t. Let $n_{hij}(t)$ be the number of years, out of N, in which day t was in state $j(j = 1$ for "rain", $j = 0$ for "dry"), given that the two preceding days $t - 1$ and $t - 2$ were in states i and h respectively; then there will be eight such numbers for day t, namely $n_{000}(t)$, $n_{001}(t)$, $n_{010}(t)$, $n_{011}(t)$, $n_{100}(t)$, $n_{101}(t)$, $n_{110}(t)$, $n_{111}(t)$. The observed proportion of rainy days on day t, for which both preceding days were dry, is therefore

$$r_{00}(t) = n_{001}(t)/\{(n_{000}(t) + n_{001}(t)\} \tag{3.9}$$

with similar definitions for $r_{01}(t)$ (for which day $t - 2$ was dry and day $t - 1$ wet); $r_{10}(t)$ (day $t - 2$ wet, day $t - 1$ dry); and $r_{11}(t)$ (both preceding days wet). We can thus calculate the four proportions $r_{hi}(t)$ for each day t of the year $(t = 1, 2, 3, \ldots)$. Note that we can also calculate the $r_{hi}(t)$ for the extra days in leap years; the difference is just that we calculate the proportions from about $N/4$ triplets, instead of the N triplets used in calculating the $r_{hi}(t)$ for the 365 remaining days.

We can plot the four curves $r_{hi}(t)$ against time, and we usually seek to model the way in which they vary throughout the year, according to some scheme which uses a small number of parameters. (Note that for a third-order Markov chain, there would be eight such curves, $r_{ghi}(t)$, corresponding to the occurrence of states g, h, and i on days $t - 3$, $t - 2$, and $t - 1$ respectively; but since we are calculating twice the number of proportions from the same N years of record, the number of quadruplets g, h, i, j used to calculate each proportion may become very small, even zero, for some days in the year.) For the second-order Markov chain, the simplistic approach would be to regress each of the four $r_{hi}(t)$ on harmonic functions of time, $\cos(2\pi t/365)$, $\sin(2\pi t/365)$; $\cos(4\pi t/365)$, $\sin(4\pi t/365)\ldots$ using standard multiple regression theory to decide when the inclusion of additional terms fails to give a significant improvement in fit; however, this would take no account of the fact that the $r_{hi}(t)$ are proportions which are constrained to lie between 0 and 1. Assuming the Markov chain model, a better approach is to take the numerator $n_{001}(t)$ in (3.9) as the number of "successes" in a Bernoulli sample of size $n_{000}(t) + n_{001}(t)$, with the probability of "success" $p_{00}(t)$. The fact that the probability $p_{00}(t)$ has lower and upper limits of 0 and 1 can be allowed for by writing

$$p_{00}(t) = \exp(g_{00}(t))/\{1 + \exp(g_{00}(t))\}, \tag{3.10}$$

where $g_{00}(t)$ is a linear function of parameters $\beta_0; \beta_1; \beta_2; \beta_3; \ldots$ of the form

$$g_{00}(t) = \beta_0 + \beta_1\cos(2\pi t/365) + \beta_2\sin(2\pi t/365) + \beta_3\cos(4\pi t/365) + \beta_4\sin(4\pi t/365) + \ldots \tag{3.11}$$

The quantity $p_{00}(t)$ in (3.10) is then a function lying between the limits 0 and 1 for all $t, t = 1, 2, \ldots 365$. Rearranging (3.10) and (3.11) we have the **logistic regression**

$$\log_e[p_{00}(t)/\{1 - p_{00}(t)\}] = \beta_0 + \beta_1\cos(2\pi t/365) + \beta_2\sin(2\pi t/365) + \beta_3\cos(4\pi t/365) + \beta_4\sin(4\pi t/365) + \ldots$$

or, in vector notation,

$$\log_e[p_{00}(t)/\{1 - p_{00}(t)\}] = \boldsymbol{\beta}^T\boldsymbol{x} \tag{3.12}$$

where $\boldsymbol{\beta}$ is a vector of coefficients, and $\boldsymbol{x}$ is a vector containing the known functions

$$1, \cos(2\pi t/365), \sin(2\pi t/365), \cos(4\pi t/365), \sin(4\pi t/365), \ldots$$

The function on the left-hand side of (3.12) is an example of a link function as defined in Chapter Two. Similar models to (3.12) are also used to represent the time dependence of $p_{01}(t), p_{10}(t)$ and $p_{11}(t)$, so that we have, in addition to (3.12), three similar relations:

$$\log_e[p_{01}(t)/\{1 - p_{01}(t)\}] = \boldsymbol{\gamma}^T\boldsymbol{x}, \tag{3.12a}$$

$$\log_e[p_{10}(t)/\{1 - p_{10}(t)\}] = \boldsymbol{\delta}^T\boldsymbol{x}, \tag{3.12b}$$

$$\log_e[p_{11}(t)/\{1 - p_{11}(t)\}] = \boldsymbol{\varphi}^T\boldsymbol{x}, \tag{3.12c}$$

where the parameters in the vectors $\boldsymbol{\beta}, \boldsymbol{\gamma}, \boldsymbol{\delta}$ and $\boldsymbol{\varphi}$ are to be estimated, using the values of $n_{hij}(t)$ obtained from the daily rainfall record, by the method described above.

The following section gives an example which (1) illustrates the estimation procedure; (2) tests whether a second-order Markov chain provides a significantly better fit than a first-order chain.

As pointed out by McCullagh and Nelder (1989), an important assumption in the model (3.12) for the occurrence of rainfall is that the probability of rain on day t depends only on whether or not it rained on day $t - 1$, but not on the amount of rain that fell, if day $t - 1$ was wet. They suggest that it may be necessary also to check whether $p_{11}(t)$ also depends on the depth of rain that fell on day $t - 1$, and suggest how this might be done; we return to this point in section 3.6.

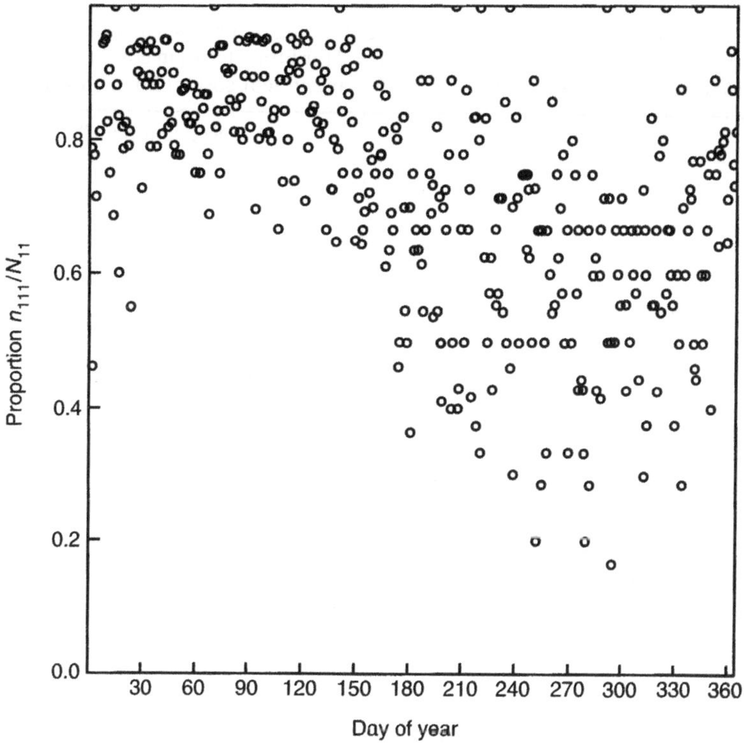

Figure 3.1 Plot of the proportions $N_{11}(t) = n_{110}(t) + n_{111}(t)$: daily rainfall data from Reserva Ducke, Manaus, Amazonas, Brazil

3.4.1 Example: daily rainfall from Reserva Ducke, Manaus, 1965–93

Reserva Ducke is a tropical forest reserve some 20 km north of the city of Manaus, State of Amazonas, Brazil. A raingauge, read daily, is sited in a forest clearing, and we consider the analysis of the occurrence of rainfall on days during the period 1965–93. Although the record is of relatively good quality, it is incomplete in that rainfall was not measured, or was not recorded, for some days during this period. We consider here the analysis of $r_{11}(t)$; the analysis for $r_{00}(t)$, $r_{01}(t)$ and $r_{10}(t)$ proceeds along similar lines.

We first look through the sequence of daily rainfalls to count the number of occurrences $n_{111}(t)$ and $n_{011}(t)$ on each day of the year, $t = 1 \ldots 365$. Using the method described in Chapter Two, we then fit the model shown in (3.12c), using the proportions $n_{111}(t)/[n_{110}(t) + n_{111}(t)]$. Figure 3.1 shows a plot of these proportions, from which a general seasonal fluctuation is apparent. The "null" model, which amounts to assuming that the probability of rain is constant throughout the year, has 364 degrees of freedom (df) gives a deviance statistic D (measuring the poorness of fit) of 670.2 ; fitting the harmonic terms

Table 3.1 Fitting the logistic regression for p_{111} (see (3.12c): values of the deviance statistic D, and degrees of freedom df, corresponding to models with different numbers of harmonics ($m = 0$ for the null model, $m = 2$ for fitting terms in $\cos(2\pi t/365)$, $\sin(2\pi t/365)$, and $m = 4$ for fitting $\cos(2\pi t/365)$, $\sin(2\pi t/365)$, $\cos(4\pi t/365)$, $\sin(4\pi t/365)$)

m	df	D	Reduction in D
0	364	670.2	
			273.8
2	362	396.4	
			2.5
4	360	393.9	

$\cos(2\pi t/365)$ and $\sin(2\pi t/365)$ reduces the df by two to 362, with deviance D reduced to 396.4. When the harmonic terms $\cos(4\pi t/365)$, $\sin(4\pi t/365)$ are also included, the df are reduced to 360 and D is further reduced to 393.9. These results can be tabulated as shown in Table 3.1.

Table 3.1 shows that the deviance statistic D is only slightly reduced when the number of explanatory variables (which are the harmonic terms) in the logistic regression (3.12c) is increased from two to four; also, this reduction is approximately distributed as a χ^2 statistic with 2 df, which has a mean value of 2. The calculated reduction in deviance, 2.5, is only slightly greater than the mean value of χ^2 calculated on the hypothesis that ϕ_3 and ϕ_4, the coefficients of the harmonic terms $\cos(4\pi t/365)$ and $\sin(4\pi t/365)$ in (3.12c), are zero. Hence the reduction in D is far from significant, and we can conclude that the two harmonic terms $\cos(2\pi t/365)$ and $\sin(2\pi t/365)$ adequately represent the seasonal variation in the probability $p_{11}(t)$.

Of course, we also need to consider the models in (3.12), (3.12a) and (3.12b). Table 3.2 shows the deviance statistics D for $m = 0$, 2 and 4 for all four models for p_{00}, p_{01}, p_{10} and p_{11}, together with the calculated coefficients β, γ, δ and ϕ. The df for $m = 1$ in Table 3.2 vary because for some days in the year, the historic record contained no appropriate sequences: thus for 25 ($= 365 - 339 - 1$) days in the year, there were no sequences in the record of the type 000 or 001; in other words these sequences were observed on $365 - 25 = 340$ days, giving the 339 degrees of freedom shown in the second column of the table.

The coefficients β_2, γ_2, δ_2 and ϕ_2 are all large relative to their approximate standard errors, and with the exceptions of β_3 and β_4, all the coefficients of terms in $\cos(4\pi t/365)$ and $\sin(4\pi t/365)$ lie within two standard deviations of their standard errors. Except for the model of p_{00}, therefore, terms in $\cos(2\pi t/365)$ and $\sin(2\pi t/365)$ describe adequately the seasonal variation in the p_{ij}. For p_{00}, this suggests that it would be worth exploring whether higher-order harmonics are also significant. When terms in $\cos(6\pi t/365)$ and $\sin(6\pi t/365)$ are fitted, the reduction in deviance is $(434.1 - 422.6) = 11.5$, with 2 df: still a significant reduction, and the coefficients of the two additional terms are 0.2689 ± 0.0827 and 0.0669 ± 0.0784, so that the term $\sin(6\pi t/365)$ predomi-

Table 3.2 Deviance statistics D, and estimated coefficients for the logistic regression models for p_{00}, p_{01}, p_{10} and p_{11}

m	df	p_{00}	df	p_{01}	df	p_{10}	df	p_{11}
0	339	518.8	361	472.1	361	509.7	364	670.2
2	337	451.6	359	402.4	359	414.2	362	396.4
4	335	434.1	357	401.3	357	411.0	360	393.9

$m = 2$			
$\beta_0 = -0.0597$	$\gamma_0 = 0.8906$	$\delta_0 = 0.3568$	$\phi_0 = 1.0580$
± 0.0601	± 0.0556	± 0.0524	± 0.0352
$\beta_1 = 0.0175$	$\gamma_1 = 0.0827$	$\delta_1 = 0.2140$	$\phi_1 = 0.1185$
± 0.0693	± 0.0738	± 0.0706	± 0.0484
$\beta_2 = 0.6808$	$\gamma_2 = 0.6403$	$\delta_2 = 0.6904$	$= 0.8269$
± 0.0805	± 0.0791	± 0.0751	± 0.0510
$m = 4$			
$\beta_0 = -0.0100$	$\gamma_0 = 0.7977$	$\delta_0 = 0.3508$	$= 1.0614$
± 0.0638	± 0.0559	± 0.0527	± 0.0355
$\beta_1 = -0.1444$	$\gamma_1 = 0.0934$	$\delta_1 = 0.1868$	$= 0.1128$
± 0.0808	± 0.0769	± 0.0721	± 0.0485
$\beta_2 = 0.8002$	$\gamma_2 = 0.6232$	$\delta_2 = 0.6930$	$= 0.8287$
± 0.0972	± 0.0805	± 0.0765	± 0.0512
$\beta_3 = -0.2056$	$\gamma_3 = 0.0737$	$\delta_3 = -0.0303$	$= -0.0189$
± 0.0810	± 0.0755	± 0.0717	± 0.0497
$\beta_4 = -0.2736$	$\gamma_4 = 0.0248$	$\delta_4 = -0.1283$	$= 0.0771$
± 0.0775	± 0.0776	± 0.0742	± 0.0499

nates. When even higher-order terms $\cos(8\pi t/365)$ and $\sin(8\pi t/365)$ are included, the reduction in deviance is $(422.6 - 417.4) = 5.2$ on 2 df, which is not significant at the 5% level. So we conclude that sine terms up to $\sin(6\pi t/365)$ are necessary to model p_{00}.

3.4.2 Example (continued): testing whether a second-order Markov chain is necessary for modelling rainfall occurrence at Reserva Ducke

The last section showed how to model the seasonal variability in p_{00}, p_{01}, p_{10} and p_{11} at Reserva Ducke. The question therefore arises: is it necessary to use a second-order Markov chain to model rainfall occurrence, or would a first-order chain have been sufficient? This problem can be approached by the inclusion of "dummy variables" in the models (3.12), (3.12a), (3.12b), (3.12c) as illustrated below.

Recall that the $p_{00}(t)$ on day t is estimated by

$$n_{001}(t)/[n_{000}(t) + n_{001}(t)],$$

with $p_{10}(t)$ estimated by

$$n_{101}(t)/[n_{100}(t) + n_{101}(t)];$$

similarly, $p_{01}(t)$ and $p_{11}(t)$ are estimated by

$$n_{011}(t)/[n_{010}(t) + n_{011}(t)] \quad \text{and} \quad n_{111}(t)/[n_{110}(t) + n_{111}(t)]$$

respectively. But if a first-order Markov chain is adequate, the occurrence or non-occurrence of rain on day $t - 2$ will convey no information about the occurrence of rain on day t: the state on day $t - 1$ will contain all useful information about the state on day t. This is equivalent to saying that

$$n_{001}(t)/[n_{000}(t) + n_{001}(t)] \quad \text{and} \quad n_{101}(t)/[n_{100}(t) + n_{101}(t)]$$

– the estimators of $p_{00}(t)$ and $p_{10}(t)$ respectively – are estimators of the same quantity, say $p_{00}(t)$. Similarly, if a first-order Markov chain is adequate,

$$n_{011}(t)/[n_{010}(t) + n_{011}(t)] \quad \text{and} \quad n_{111}(t)/[n_{110}(t) + n_{111}(t)]$$

– the estimators of $p_{01}(t)$ and $p_{11}(t)$ respectively – will also estimate the same quantity, $p_{11}(t)$ say. Thus, we need to test whether the harmonic curves fitted to $p_{00}(t)$ and $p_{10}(t)$ differ significantly, and whether the harmonic curves fitted to $p_{01}(t)$ and $p_{11}(t)$ differ significantly; if the two curves in each pair do not differ significantly one from the other, we can conclude that a first-order Markov is adequate to represent the occurrence of rainfall at Reserva Ducke.

The procedure for comparing two fitted curves is similar to that used for comparing two linear regressions, when the test procedure falls into two parts. First, the slopes of the two linear regressions are compared; if they differ, we conclude that the two regressions cannot be combined, and the testing procedure stops. But if the slopes of the two regressions do not differ by more than would be expected by chance, the two estimates of the common slope are combined to form a pooled estimate. However, the two regressions may then still differ in their intercepts, even though the preceding test has shown that they do not depart from parallelism; hence the next and final step is to test whether the parallel regressions differ significantly in their intercepts. If they do not, a pooled estimate of the common intercept is calculated.

A similar procedure is used for comparing the two harmonic curves for $p_{00}(t)$ and $p_{10}(t)$ (and for $p_{01}(t)$ and $p_{11}(t)$ also). First, we test for parallelism; if the null hypothesis that the curves are parallel is not rejected, pooled estimates of the model coefficients are calculated, and the curves are tested for differences between their intercepts. If the null hypothesis of equal intercepts is not rejected, a pooled estimate of the intercept (the model constant) is calculated. If either null hypothesis is rejected, it is concluded that the curves cannot be combined: if, in the present example, the curves in either pair cannot be combined, we conclude that a first-order Markov chain is not appropriate, and that a second-order chain is required.

To explain the procedure, suppose that the fitted curves had just two harmonic terms $\cos(2\pi t/365)$ and $\sin(2\pi t/365)$, which we denote by c_t and s_t, so that the curves fitted for $p_{01}(t)$ and $p_{11}(t)$ are

$$\log_e[p_{01}(t)/\{1 - p_{01}(t)\}] = \gamma_0 + \gamma_1 c_t + \gamma_2 s_t \qquad (3.13a)$$

and

$$\log_e[p_{11}(t)/\{1 - p_{11}(t)\}] = \phi_0 + \phi_1 c_t + \phi_2 s_t. \tag{3.13b}$$

We now put all the 730 proportions

$$n_{k11}(t)/[n_{k10}(t) + n_{k11}(t)] \quad \text{for } k = 0, 1$$

together and construct some new or "dummy" variables, u_t, v_t and w_t say, defined by

$$\begin{aligned} u_t &= 0 \quad \text{for } k = 0 \\ &= 1 \quad \text{for } k = 1 \end{aligned}$$

$$\begin{aligned} v_t &= 0 \quad \text{for } k = 0 \\ &= c_t \quad \text{for } k = 1 \end{aligned}$$

and

$$\begin{aligned} w_t &= 0 \quad \text{for } k = 0 \\ &= s_t \quad \text{for } k = 1 \end{aligned}$$

and fit the model

$$\begin{aligned} \log_e[p_{k1}(t)/\{1 - p_{k1}(t)\}] &= \theta_0 + \theta_1 u_t + \theta_2 c_t + \theta_3 s_t \\ &\quad + \theta_4 v_t + \theta_5 w_t \quad \text{for } k = 0, 1. \end{aligned}$$

For $k = 0$ the model becomes

$$\log_e[p_{01}(t)/\{1 - p_{01}(t)\}] = \theta_0 + \theta_2 c_t + \theta_3 s_t \tag{3.14a}$$

(since u_t, v_t and w_t are defined to be zero for $k = 0$), whilst for $k = 1$, the model becomes

$$\begin{aligned} \log_e[p_{11}(t)/\{1 - p_{11}(t)\}] &= \theta_0 + \theta_1 u_t + \theta_2 c_t + \theta_3 s_t + \theta_4 v_t + \theta_5 w_t \\ &= (\theta_0 + \theta_1) + (\theta_1 + \theta_3)c_t + (\theta_2 + \theta_4)s_t \end{aligned} \tag{3.14b}$$

since u_t, v_t and w_t are defined to be 1, c_t and s_t for $k = 1$. Comparing (3.14a) with (3.14b), we see that a test of parallelism is obtained by testing the null hypothesis $H_0\!: \theta_3 = \theta_4 = 0$ against the alternative $H_1\!: \theta_3 \neq \theta_4 \neq 0$. The test used is a likelihood ratio test as described in Chapter Two. If the data prove to be consistent with H_0, we proceed to test $H_0\!: \theta_1 = 0$ against the alternative $H_1\!: \theta_0 \neq 0$; if the data are then consistent with H_0, we are justified in using the same curve for both p_{01} and p_{11}, say

$$\log_e[p_1(t)/\{1 - p_1(t)\}] = \theta_0 + \theta_2 c_t + \theta_3 s_t. \tag{3.15}$$

The argument is essentially the same when the right-hand side of (3.15) contains terms in $\cos(4\pi t/365)$ and $\sin(4\pi t/365), \ldots$; but it then becomes necessary to introduce more dummy variables.

We now look at how dummy variables are used to test for parallelism in

Table 3.3 Analysis of deviance statistic D, for testing parallelism of models (3.13a) and (3.13b) for p_{01} and p_{11}. Variables c_{2t}, s_{2t} denote $\cos(2\pi t/365), \sin(2\pi t/365)$; similarly c_{4t} and s_{4t} denote $\cos(4\pi t/365), \sin(4\pi t/365)$. The dummy variables fitted are $u_t c_{2t}, u_t s_{2t}, u_t c_{4t}$, and $u_t s_{4t}$ where $u_t = 0$ for $k = 0$ and $u_t = 1$ for $k = 1$

Deviance due to	df	D	Changes in df and D	
Fitting "null" model	726	1209.4		
			4	397.0
Fitting terms $c_{2t}, s_{2t}, c_{4t}, s_{4t}$	722	812.4		
			4	1.7
Further inclusion of dummy variables	718	810.7		

the fitted models for $\log_e[p_{01}(t)/\{1 - p_{01}(t)\}]$ (see 3.13a) and $\log_e[p_{11}(t)/\{1 - p_{11}(t)\}]$ (see 3.13b). The analysis of deviance is shown in Table 3.3.

The inclusion of the four dummy variables therefore reduces the deviance statistic D by only 1.7; testing this as an approximate χ^2 with 4 df shows that the deviance reduction is not significant. We can conclude that there is no evidence that the coefficients of the sine and cosine terms in the models for p_{01} and p_{11} differ significantly, and that the fitted curves are parallel.

The next stage, therefore, is to test whether the two curves differ significantly in their displacement. We might get a hint of the result from inspection of Table 3.2, where, for $m = 2$, the constant term γ_0 in the model for p_{01} is estimated as 0.8906 ± 0.0556, and the constant term ϕ_0 in the model for p_{11} is estimated as 1.0580 ± 0.0352; the difference between the two constants is 0.1674 ± 0.0658, rather greater than twice its standard error. When the test is effected in terms of the deviance statistic, this hint is confirmed, as Table 3.4 shows.

Table 3.4 shows that the reduction in the deviance statistic that results from the inclusion of a term for differences between the two constant terms γ_0 and ϕ_0 (equivalent to testing the null hypothesis $H_0: \theta_1 = 0$ in (3.14b)) is 12.1 with 1 df; this is tested as an approximate χ^2 with 1 df, which indicates strong evidence against the null hypothesis ($P < 0.001$). We conclude that, although the fitted curves in the models (3.12a) and (3.12c) for p_{01} and p_{11} can be taken as parallel, they differ significantly in their displacement: hence the two probabilities p_{01} and p_{11} cannot be replaced by a single probability p_1, as a first-order Markov chain would require.

We have restricted the discussion to an examination of whether p_{01} and p_{11} can be combined, and have shown that they cannot. A similar argument, not presented here, confirms that p_{00} and p_{10} cannot be combined either, as might have been expected from consideration of Table 3.2. We are led to the conclusion that the occurrence of daily rainfall at Reserva Ducke, Manaus, cannot be simply represented by a first-order Markov chain. The example has not, of course, tested whether a second-order chain is sufficient, or whether a

Table 3.4 Test for difference in displacement between the two parallel curves fitted for p_{01} and p_{11}

Deviance due to	df	D	Changes in df and D	
Fitting "null" model	726	1209.4		
			2	394.5
Fitting terms c_{2t}, s_{2t}	724	814.9		
			1	12.1
Fitting term for differences in γ_0 and ϕ_0	723	802.8		

higher-order chain is suggested by the data; the test for this follows essentially the same lines as those used in this example. Stern and Coe (1984), whose methods are the basis of this chapter, report:

> At some sites second-order chains are required. However, models with three rather than four curves may prove adequate with the probability of rain depending only on the state of the previous day if it was rainy. This idea may be extended to models with order greater than two. At Nagpur (India) . . . the model finally adopted was a fifth-order chain, which required only the six curves $p_1(t), p_{10}(t) \ldots p_{10000}(t)$ and $p_{00000}(t)$, where, for example, $p_{100}(t)$ = probability of rain following exactly two dry days. The further constraint that all the curves except $p_1(t)$ were parallel on the logit scale also proved to be reasonable.

An important point, however, is that where high-order chains are to be fitted, the number of years of daily rainfall commonly available in many developing countries results in numbers of binomial trials that are often very small and sometimes zero; discretion may then be required to find a balance between the limited availability of data and the need for greater model complexity.

3.5 THE USE OF MODELS OF RAINFALL OCCURRENCE TO GIVE ANSWERS TO QUESTIONS OF PRACTICAL INTEREST

Fitting a model of the type considered above is not, of course, an end in itself; the model is required to help answer questions of practical importance in, for example, agricultural planning. Stern and Coe (1982) identify three methods of obtaining useful results from such a model; they are as follows:

1. **Analytic**, in which a formula is derived to give the required result in terms of the model parameters. In practice this is very difficult for all but the simplest of models.
2. By the use of **recurrence relations** based upon the Markov properties. This method can be used, for example, to calculate the probability distribution of the time of onset of the rainy period at a site with strongly defined wet and dry seasons; and to calculate the probability of long dry spells begin-

ning at any date in the growing season of a crop.

3. By **computer simulation**. The fitted models are used to generate long artificial sequences of rainfall occurrence, which are then analysed to obtain estimates of how frequently an event of interest occurs. This method is also easily extended to the case where the quantities of rain falling on wet days are also to be modelled: a matter not yet considered in this chapter. For example, in seasonally arid countries, the start of the growing season has been defined (Virmani, 1975, cited by Stern and Coe, 1982) as occurring when 20 mm of rain falling on one or two successive days; each year of simulated data will yield a date on which this event occur, and these dates will yield a histogram for the start of growth.

Later we shall illustrate the use of method (2) using the daily rainfall sequence from Reserva Ducke, Manaus. First, however, it is necessary to treat the other aspect of modelling daily rainfall: namely, the **modelling of rainfall depths** on wet days.

3.6 THE MODELLING OF RAINFALL DEPTHS

This section pursues further the methods described by Stern and Coe (1984). A model for rainfall depths must describe the probability distribution of amount of rain, and the way in which this distribution varies as t varies thoughout the year and, possibly, on what has occurred on the preceding days $t-1, t-2, \ldots$. Denote the occurrence of rain by the random variable $J(t)$ with $J(t) = 0$ on dry days, $J(t) = 1$ on wet days, and let $X(t)$ be the depth of rain on day t when $J(t) = 1$. In practice, since there is lower limit to the depth of rain that is measured – such as 0.1 mm, or even some greater value – the variable $X(t)$ is defined as the rainfall in excess of this lower limit. Inspection of rainfall depths on any given day t shows a distribution that is almost always highly skewed, and gamma distributions are commonly used to describe them. A two-parameter form of this distribution is

$$f(x) = [\kappa/\mu(t)]^{\kappa} x^{\kappa-1} \exp[-\kappa x/\mu(t)]/\Gamma(\kappa), \tag{3.16}$$

where the time variation in the parameter $\mu = \mu(t)$ has been indicated, and κ is the "shape" parameter κ which determines the coefficient of variation $1/\sqrt{\kappa}$ of the gamma distribution. It is commonly valid to assume that κ is constant; when $\kappa = 1$ the distribution in (3.16) reduces to the exponential distribution, as mentioned in Chapter Two. Explicitly, the time dependence in $\mu(t)$ is commonly represented by link functions of the form

$$\log_e[\mu(t)] = \lambda_0 + \lambda_1 \cos(2\pi t/365) + \lambda_2 \sin(2\pi t/365) + \lambda_3 \cos(4\pi t/365) + \lambda_4 \sin(4\pi t/365) + \ldots \tag{3.17}$$

As with the probabilities p_{00}, p_{01}, p_{10} and p_{11} for the occurrence of rain in the Reserva Ducke example, a primary concern is the number of harmonics necessary to describe the time variation in $\mu(t)$ throughout the year. As before, models with increasing numbers of harmonics are fitted by maximum likelihood, and the deviance statistic D is calculated for each; we take as the form of the fitted model that for which the inclusion of additional harmonic terms results in no significant reduction in D.

Suppose that a second-order Markov chain is used for describing the occurrence of rainfall; then we need to consider the time variation in mean daily rainfall on wet days (state 1 on day t) that follow (1) two dry days (state 0 on days $t-1$ and day $t-2$); (2) a dry day (state 0 on day $t-1$) following a wet day (state 1 on day $t-2$); (3) a wet day (state 1 on day $t-1$) following a dry day (state 0 on day $t-2$); and finally (4) two wet days (state 1 on days $t-1$ and $t-2$). Denoting the four sequences of states by the suffices h, i, j, we therefore have to deal with four sets of means $\mu_{hi}(t), t = 1 \ldots 365$, and the suffices (hi) take values (00), (10), (01) and (11) in the four cases (1) to (4) listed above. Clearly, since we are dealing with days with rain, $j = 1$ in all four cases.

To fix ideas, suppose that we wish to model the time dependence of $\mu_{11}(t)$, case (4) above; the argument is the same for $\mu_{00}(t)$, $\mu_{01}(t)$ and $\mu_{10}(t)$. For each day t of the year ($t = 1, 2 \ldots 365$), the historic record is scanned to determine how many times it rained on day and on the two preceding days also ($h = 1, i = 1, j = 1$); this number is denoted by $n(t)$. The $n(t)$ rainfalls are also extracted, and their sums $\Sigma x(t)$ and $\Sigma \log_e x(t)$ calculated together with the mean values of these quantities. Thus we have a sequence of 365 mean values for $\bar{x}$ (the mean rainfall on wet days that are also preceded by two wet days), which are used to fit the model (3.17). The values obtained for the mean values $\bar{x}$ are shown in Figure 3.2; once again, a seasonal fluctuation is apparent. In the terminology of Chapter Two, a generalised linear model (GLM) is fitted with the data gamma-distributed, with a log-link function and with the linear predictor containing explanatory variables which are harmonic functions of time.

In section 3.4 was mentioned the suggestion of McCullagh and Nelder (1989) that the probability of rain on day t, given that day $t-1$ was also wet, may depend upon the depth of rain that fell on day $t-1$, and not only on the fact that day $t-1$ was wet. They suggest that this possibility should be dealt with by incorporating the depth of rainfall (in log units) in the model for $p_{11}(t)$ given in (3.12).

3.6.1 Example (continued): fitting harmonics to the mean rain per wet day, on days following two previous days with rain: Reserva Ducke, Manaus

As an example, we consider the analysis of deviance for testing the effects of increasing the number of harmonics fitted to $\bar{x}$, mean rainfall on days preceded by two wet days. The model fitted is

$$\mathrm{E}[X] = \mu_{11}(t)$$

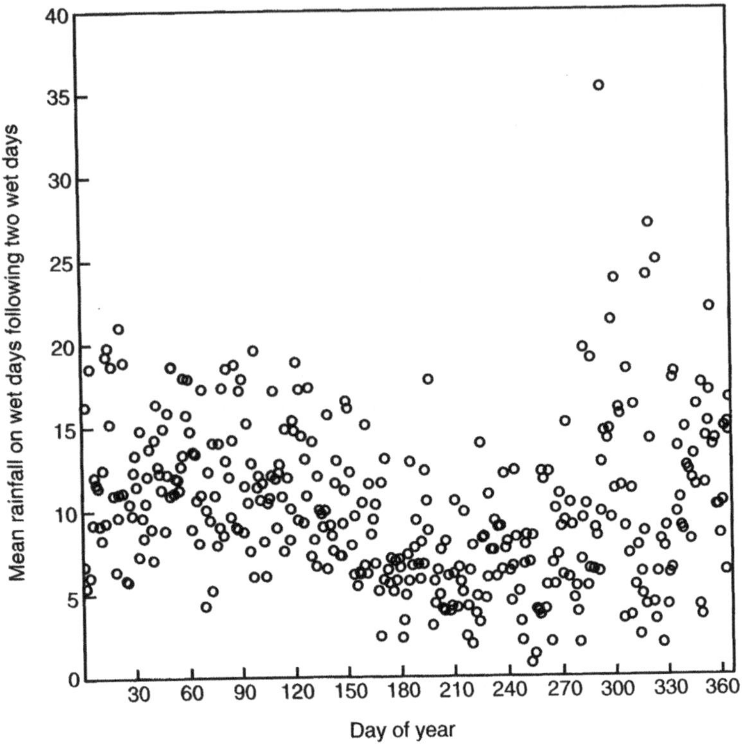

Figure 3.2 Mean values of rainfall on wet days that are also preceded by two wet days

with the random variable X gamma-distributed;

$$\log_e[\mu_{11}(t)] = \eta(t),$$

where the linear predictor $\eta(t)$ has the assumed form

$$\eta(t) = \lambda_0 + \lambda_1\cos(2\pi t/365) + \lambda_2\sin(2\pi t/365) + \lambda_3\cos(4\pi t/365) + \lambda_4\sin(4\pi t/365) + \ldots$$

The objective is to determine how many harmonics are required to model $\mu_{11}(t)$.

The mean values $\bar{x}$ described in the last section are shown in Figure 3.2, but the number of rainfalls used to calculate each mean varies with t. Therefore, we need to use the numbers of rainfalls $n(t)$ as weights when fitting the model.

Table 3.5 shows that inclusion of terms $\cos(6\pi t/365)$, $\sin(6\pi t/365)$ results in a small reduction of deviance, 1.9 on 2 df. As an approximate test, this value is compared with tabulated values of χ^2; it is clearly far from significant, being

Table 3.5 Analysis of deviance for testing the effect of increasing the number of harmonics used to model $\mu_{11}(t)$

Deviance due to	df	D	Changes in df and D	
Fitting the "null" model	364	798.4		
			2	91.5
Fitting $\cos(2\pi t/365), \sin(2\pi t/365)$	362	707.0		
			2	16.4
Fitting $\cos(4\pi t/365), \sin(4\pi t/365)$	360	690.6		
			2	1.9
Fitting $\cos(6\pi t/365), \sin(6\pi t/365)$	358	688.7		

close to the mean value of the χ^2 distribution. We can conclude that $m = 4$ harmonics – terms in $\cos(2\pi t/365), \sin(2\pi t/365)$ and $\cos(4\pi t/365), \sin(4\pi t/365)$ – are required to model the time variation in $\mu_{11}(t)$.

Section 3.4.2 showed how dummy variables are used to test whether fitted harmonic curves are parallel, and if so, whether they differ in displacement. When modelling the $\mu_{hi}(t)$ of a second-order Markov process, we also seek to reduce the number of parameters necessary to model them by comparing whether the four curves $\mu_{hi}(t)(h, i = 0, 1)$ can be reduced to some smaller number, and the testing procedure follows the same pattern: dummy variables are introduced, and the test for parallelism is made by testing the null hypothesis that the coefficients of the dummy variables are zero. If this hypothesis is not rejected, pooled coefficients for the four curves are fitted, and the hypothesis that the four curves have the same constant term (corresponding to the $\beta_0, \gamma_0, \delta_0, \phi_0$ of section 3.4.2) follows.

3.7 ESTIMATION OF THE SHAPE PARAMETER κ OF THE GAMMA DISTRIBUTIONS FOR RAINFALL DEPTHS

The remaining problem in the modelling of rainfall occurrences and depths lies in the estimation of the shape parameter κ of the gamma distribution of rainfall depths. Two estimates are possible, according to whether or not the inaccuracies in the estimation of small rainfall depths are important or can be ignored.

Where these inaccuracies can be ignored, Stern and Coe (1984) recommend that κ be estimated from the **within-day deviance** calculated as

$$D = 2 \sum_{t=1}^{T} n(t)[\log_e \bar{x}(t) - [\log_e x(t)]],$$

where

$$[\log_e x(t)] = \sum_{t=1}^{T} \log_e[x(t)]/n(t).$$

A maximum likelihood estimate of κ is then obtained by solving the equation

$$\log_e \kappa - \psi(\kappa) = D/2(n - T), \tag{3.19}$$

where $n = \Sigma n(t)$, $\psi(\kappa)$ is the digamma function (Abrahamowitz and Stegun, 1968) which is now available in several statistical packages, and T is the number of days in the period being modelled: T is 365 in the present case where we are modelling the whole year, but T would be less if, say, we were modelling just a part of the year (e.g. $T = 92$ for the period October to December). An approximate solution to (3.19) is

$$\hat{\kappa} = (n - T)/D. \tag{3.20}$$

Where the errors in measurement of small rainfall depths need to be taken into account, Stern and Coe (1984) give an alternative estimate of κ based upon the **residual between-day deviance** D_b. A maximum likelihood estimate of κ is then obtained by solving for κ the equation

$$\sum_{t=1}^{T} n(t)[\log_e(n(t)\kappa) - \psi(n(t)\kappa)] = D_b/2 \tag{3.21}$$

If $n(t)\kappa$ is greater than about 10, an approximate solution is T/D_b.

3.7.1 Example (continued): estimation of κ from daily rainfall data at Reserva Ducke, Manaus

We illustrate the estimation of κ corresponding to the model of $\mu_{11}(t)$ obtained in the previous section. The tropical rainfall regime at Reserva Ducke, Manaus, does not suggest that errors in the measurement of small rainfall depths are likely to be important so we use (3.18) to (3.20). Using the sequence $n(t)$ obtained for that term, the within-day deviance calculated from the $n = 3742$ rainfalls and (3.18) is 6345.1, based upon 3741 degrees of freedom. Hence the approximate solution of (3.19), given by (3.20), is

$$\hat{\kappa} = (3742 - 365)/6345.1 = 0.5322.$$

The question arises, how far is it justifiable to assume that κ is constant for each day of the year? To look at this question, an estimate of κ was calculated for each day by solving the equation (3.19) iteratively using the numerical procedure of Greenwood and Durand (1960). The estimates obtained are plotted against time in Figure 3.3. Some of the estimates of κ were, of course, calculated from very small samples of data, and Figure 3.4 shows the numbers of data values used for each estimate. It is clear that the more divergent estimates of κ are obtained during the "dry" season when sequences of two days with rain are much less frequent. The mean estimate of all the κ, weighted with the number of data values, is 0.7714, rather greater than the estimate of 0.5322 obtained above. However, the daily estimates were not adjusted for

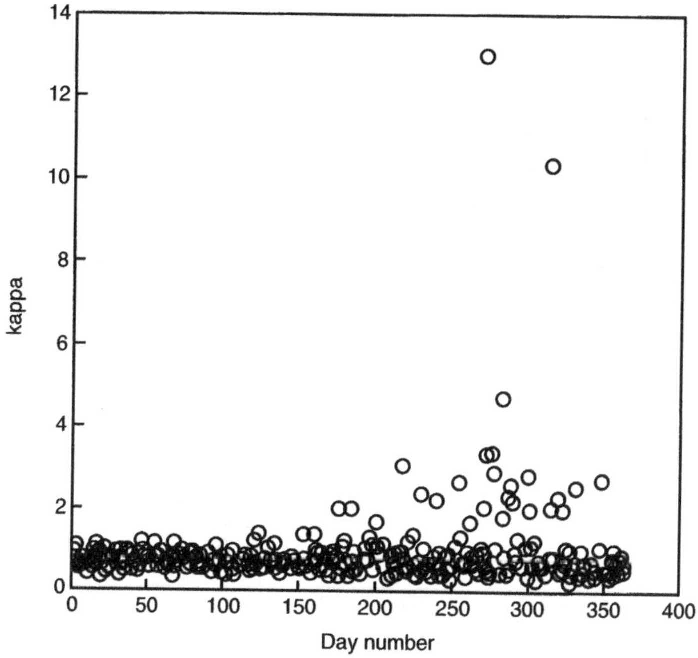

Figure 3.3 Estimates of gamma shape parameter κ ('kappa'), calculated for each day, plotted against day number ($t = 1 \ldots 365$)

bias which is known to exist; it is interesting to compare the two estimates obtained here, 0.7714 and 0.5322 with those quoted by Stern and Coe (1984); for the rainfall record at Morogoro in Tanzania, the quote is a maximum likelihood estimate of 0.768, together with a bias-reduced estimate of 0.521.

For Reserva Ducke, estimates of κ were obtained on 354 days; on the remaining 12, the number of data values $n(t)$ was too small to yield an estimate (for example, with $n(t) = 1$ it is clearly not possible to estimate the shape parameter of the gamma distribution). Of the 354 estimates of κ, 290 were less than one, suggesting that the simpler hypothesis that rainfall depths are exponentially distributed (i.e. $\kappa = 1$) would not be justified for daily rainfall at Reserva Ducke, Manaus. For any particular day, it is possible formally to test the hypothesis H_0: $\kappa = 1$ against the alternative hypothesis H_1: $\kappa \neq 1$ by using a likelihood ratio test: if n rainfall values were recorded on the day in question, with mean $\bar{x}$ and κ estimated by $\hat{\kappa}$, the maximised likelihood under H_0 is

$$\log_e L_{\max}(0) = -n([\log_e x] + 1) \tag{3.22}$$

and the maximised likelihood under the alternative hypothesis H_1 is

$$\log_e L_{\max}(1) = n\hat{\kappa}\log_e(\hat{\kappa}/\bar{x}) + n(\hat{\kappa} - 1)(\log_e x) - n\hat{\kappa} - n\log_e[\Gamma(\hat{\kappa})]. \tag{3.23}$$

Then when n is sufficiently large, the statistic

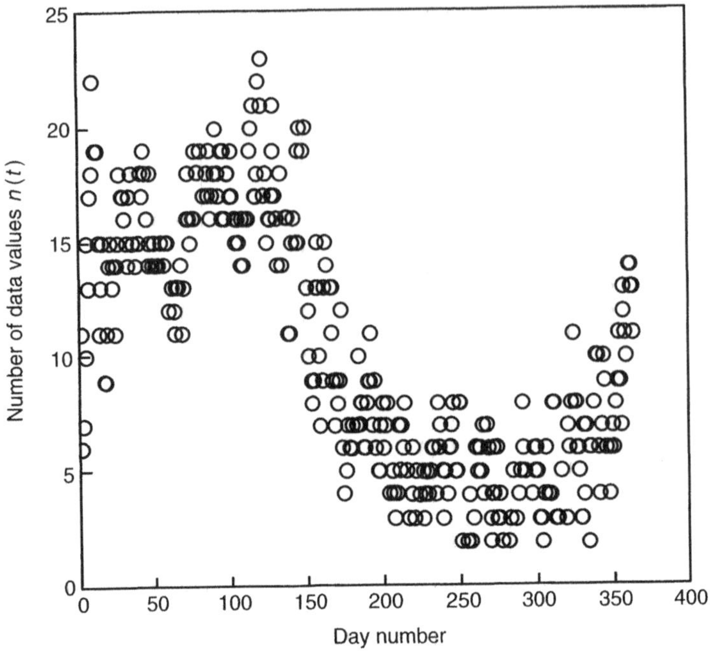

Figure 3.4 Numbers of daily rainfalls used to calculate estimates of gamma shape parameter κ, plotted against day number ($t = 1 \ldots 365$)

$$R = -2[\log_e L_{max}(0) - \log_e L_{max}(1)]$$

has a distribution that is approximately χ^2 with 1 df; tables show that the critical values for significance probabilities 0.05, 0.01 and 0.01 are 3.841, 6.635 and 10.827. Hence H_0 is accepted if R is smaller than 3.841.

3.8 USE OF THE MODEL: RECURRENCE RELATIONS

Fitting models to data becomes a sterile exercise unless the fitted model can be used to answer questions of practical interest. Section 3.5 described three procedures by which models of the kind discussed in this chapter can be used for these purposes, and in this section we concentrate on the second of those procedures described, like much of the material presented in this chapter, by Stern and Coe (1984): namely the use of recurrence relations. We now have a model for the occurrence of rain, denoted by the random variable $J(t)$ with two states 0 and 1; and a second random variable for the depth of rain on a wet day. So we must consider the **bivariate Markov chain** defined by the pair $(Y(t), J(t))$, where the states of $Y(t)$ will depend on the particular rainfall characteristic whose probability is being calculated. The states of $Y(t)$ are ordered, and the first and last states are barriers which may be **absorbing barriers** or **reflecting barriers**. We deal first with the case where $J(t)$ is a first-order Markov chain. Although it is possible (Lloyd, 1979) to reduce a bivariate chain to a univariate chain, Stern and Coe (1984) show that it is often computationally convenient to keep the bivariate form. Using the theorem of total probability (see Chapter Two), we can write

$$P[J(t)=j,\, Y(t)=l] = \sum_k \sum_i P[J(t) = j,\, Y(t)=l \mid J(t-1)=i,\, Y(t-1)=k].P[J(t-1)=i,\, Y(t-1)=k] \quad (3.24)$$

or

$$P[J(t)=j,\, Y(t)=l] = \sum \sum s_{ijkl} P[J(t-1)=i,\, Y(t-1)=k] \quad (3.25)$$

where

$$s_{ijkl} = P[J(t)=j,\, Y(t)=l \mid J(t-1)=i,\, Y(t-1)=k]. \quad (3.26)$$

But transitions between the two states of the random variable $J(t)$ depend only on the transition probabilities

$$p_{01}(t) = P[J(t)=1 \mid J(t-1)=0] \quad \text{and} \quad p_{11}(t) = P[J(t)=1 \mid J(t-1)=1].$$

Also, by the definition of conditional probability, we have

$$\begin{aligned} s_{ijkl} &= P[J(t)=j,\, Y(t)=l \mid J(t-1)=i,\, Y(t-1)=k] \\ &= P[Y(t)=l \mid Y(t-1)=k,\, J(t)=j,\, J(t-1)=i].P[J(t)=j \; J(t-1)=i,\, Y(t-1)=k] \\ &= q_{ijkl}(t).P[J(t)=j \mid J(t-1)=i,\, Y(t-1)=k], \text{ say.} \end{aligned} \quad (3.27)$$

But the second probability,

$$P[J(t)=j \mid J(t-1)=i,\, Y(t-1)=k],$$

is simply $P[J(t)=j \mid J(t-1)=i]$ since the state of $J(t)$ depends only upon the state of $J(t-1)$ and therefore is equal to the probability $p_{ij}(t)$ of transition from state i to state j. For the particular cases $j=0$ and $j=1$, corresponding to dry and wet states on the tth day, we have

$$\left.\begin{aligned} s_{i0kl}(t) &= p_{i0}(t).q_{i0kl}(t) \\ s_{i1kl}(t) &= p_{i1}(t).q_{i1kl}(t) \end{aligned}\right\} \text{for } i = 0, 1. \quad (3.28)$$

The values of the $q_{ijkl}(t)$ depend upon the particular rainfall event being studied, and Stern and Coe (1984) give three examples, presented in the following three sections.

3.8.1 Example: calculating the probability distribution of the number of wet days in a period

Suppose that the period in question is the sequence of m days beginning after time t_0, consisting of the days $t_0 + 1, t_0 + 2, \ldots t_0 + m$. The random variable $Y(t)$ is then the number of wet days in this period, so that the possible states of $Y(t)$ are

$$Y(t) = 0,\ Y(t) = 1, \ldots\ Y(t) = m.$$

Recall from the previous section that

$$q_{ijkl}(t) = P[Y(t) = l \mid Y(t-1) = k, J(t) = j, J(t-1) = i]$$

and it is clear that many of the $q_{ijkl}(t)$ are zero; for example, if $l < k, q_{ijkl}(t)$ will always be zero since the number of wet days can never decrease from one day to the next within the m-day period. Similarly, if $l > k, q_{ijkl}(t)$ will be zero whenever $J(t) = 0$. In fact the only $q_{ijkl}(t)$ that are not zero are $q_{i0kk}(t)$ – which must have value 1, since a transition to $J(t) = 0$ on day t must leave unchanged the number of wet days accumulated up to that day, whatever the state of $J(t-1)$ – and $q_{i1k,k+1}(t)$, which must also have value 1, since a transition to state $J(t) = 1$ on day t must always increase by one the number of wet days accumulated up to that day. Now (3.25) is a recurrence relation which expresses $P[J(t) = j, Y(t) = l]$ in terms of $P[J(t-1) = i, Y(t-1) = k]$, and the coefficients s_{ijkl} in this recurrence relation are known, being expressed in terms of the known quantities $q_{ijkl}(t)$ – most of which, as we have seen, are zero – and the transition probabilities $p_{ij}(t)$ (see (3.28)). We need, however, to define an initial condition, and this is taken by allocating values to the following table of probabilities at time t_0:

$P[Y(t_0) = 0, J(t_0) = 0]$	$P[Y(t_0) = 1, J(t_0) = 0]$	...	$P[Y(t_0) = m, J(t_0) = 0]$	$P[J(t_0) = 0]$
$P[Y(t_0) = 0, J(t_0) = 1]$	$P[Y(t_0) = 1, J(t_0) = 1]$	...	$P[Y(t_0) = m, J(t_0) = 1]$	$P[J(t_0) = 1]$
$P[Y(t_0) = 0]$	$P[Y(t_0) = 1]$	...	$P[Y(t_0) = m]$	1

If we choose, say, to count the number of wet days with day t_0 being dry, the only non-zero value in the body of this table is that in the top left-hand corner, for which

$$P[Y(t_0) = 0, J(t_0) = 0] = 1;$$

in the margins of the table,

$$P[J(t_0) = 0] = 1 \quad \text{and} \quad P[Y(t_0) = 0] = 1,$$

all other marginal entries being zero. Alternatively, we may wish to take day t_0 as a wet day, in which case

$$P[Y(t_0) = 0, J(t_0) = 1] = 1, P[J(t_0) = 1] \quad \text{and} \quad P[Y(t_0) = 0] = 1,$$

the remaining entries in the table being zero.

The above discussion has dealt with the case where the occurrence of rainfall was modelled by a first-order Markov chain. The recurrence relations are considerably more complicated where chains of higher order than the first are used, but for some applications (Stern and Coe, 1982) it has been found that the use of recurrence relations took about one-twentieth of the computer time required when the fitted model was used to get the same result by generating artificial rainfall sequences. A further advantage is the ease with which initial conditions can be changed and the calculation restarted. However, computing facilities have developed greatly since the pioneering work of those authors, and computing time is now rarely a significant constraint. Furthermore, where the random variable $Y(t)$ describes the occurrence of a complicated event, it is likely that direct simulation will be the more efficient approach when account is taken of user time.

As an example, we consider again the record of daily rainfall from Reserva Ducke, Manaus. To simplify matters, we assume:

1. That a first-order, two-state Markov chain adequately represents the occurrence of rainfall (although arguments presented above suggest that a first-order chain is inadequate);
2. That two harmonic terms are sufficient to represent the time dependence in the transition probabilities.

Using the Reserva Ducke record, it can be shown that the model for the occurrence of rainfall is then

$$\log_e[p_{01}(t)/(1 - p_{01}(t))] = \underset{(\pm 0.0385)}{0.1551} + \underset{(\pm 0.0489)}{0.1584}\cos(2\pi t/365) + \underset{(\pm 0.0545)}{0.7292}\sin(2\pi t/365) \quad (3.31a)$$

$$\log_e[p_{11}(t)/(1 - p_{11}(t))] = \underset{(\pm 0.0290)}{0.9976} + \underset{(\pm 0.0403)}{0.1612}\cos(2\pi t/365) + \underset{(\pm 0.0418)}{0.8050}\sin(2\pi t/365) \quad (3.31b)$$

whence, for example, the estimated transition probability $p_{01}(t)$ can be found for any day t as $p_{01}(t) = A/(1 + A)$, where

$$A = \exp[0.1551 + 0.1584\cos(2\pi t/365) + 0.7292\sin(2\pi t/365)].$$

The estimate of $p_{00}(t) = 1 - p_{01}(t)$; similarly, we find $p_{10}(t)$ and $p_{11}(t)$. We divide the year into, say, 10-day periods; then use of the recurrence relation (3.24), with the initial condition that day t_0 is dry, yields the probability distribution of the number of wet days in each of the 10-day periods. Table 3.6

shows the 36 discrete pdfs obtained; Figure 3.5a and 3.5b shows plots of the means and standard deviations of these distributions. A different initial condition for day t_0 (i.e. day t_0 wet) would yield different distributions.

3.8.2 Example: calculating the probability of a dry spell of K days or more, during the m-day period $t_0 + 1, t_0 + 2, \ldots t_0 + m$

A second example of the use of the recurrence relations (3.24) considers the probability of a dry spell of K days or more within the m-day period (typically, this period will be the critical period after emergence of a seedling, when a dry spell can be particularly damaging). Again, we assume the use of a first-order chain for simplicity.

In this application, the states of the random variable $Y(t)$ are $1, 2, \ldots K$. The state K is an **absorbing barrier** in that when $Y(t)$ enters this state, a K-day dry spell will necessarily have occurred during the m-day period: the occurrence of further dry and wet days (i.e. the later states of $J(t)$, after the K-day dry spell has occurred) cannot "undo" the K-day dry spell.

Definition of the quantities $q_{ijkl}(t)$ requires a little more care than in the preceding example. Provided that a K-day dry spell has not yet occurred as time t advances through the days $t_0 + 1, t_0 + 2, \ldots t_0 + m$, a transition from state $J(t-1) = 0$ to $J(t) = 0$ (dry day following dry day) will obviously increase the current length of dry period, k days say, by one day, with probability one; hence we have

$$q_{00k(k+1)}(t) = 1 \quad \text{for } k < K. \tag{3.32}$$

Also, if a wet day (state $J(t) = 1$) follows a dry day (state $J(t-1) = 0$) the current length of dry period, k days, ends, and we go back to the beginning to count the length of dry spell; hence,

$$q_{01k1}(t) = 1 \quad \text{for } k < K. \tag{3.33}$$

Again, if the preceding day was wet, (state $J(t-1) = 1$) the random variable $Y(t)$ will remain in its lowest state $Y(t) = 1$, whether or not it rains on day t: hence

$$q_{1011}(t) = 1 \quad \text{and} \quad q_{1111}(t) = 1. \tag{3.34}$$

Finally, whether or not it rained on the preceding day (state $J(t-1) = 0$ or state $J(t-1) = 1$), and whether or not it rains on day t (state $J(t) = 0$ or state $J(t) = 1$), if a K-day dry spell has already occurred during the m-day period, $Y(t)$ will remain in its highest state $Y(t) = K$; hence

$$q_{00KK} = 1; q_{01KK} = 1; q_{10KK} = 1; \quad \text{and} \quad q_{11KK} = 1. \tag{3.35}$$

Table 3.6 Probability distributions of number N of wet days ($N = 0, 1, \dots 10$) in consecutive 10-day periods: Reserva Ducke, Manaus, Amazonia

					N:					
0	1	2	3	4	5	6	7	8	9	10
0.00	0.00	0.01	0.02	0.05	0.10	0.17	0.22	0.22	0.15	0.06
0.00	0.00	0.00	0.01	0.03	0.08	0.15	0.22	0.24	0.19	0.08
0.00	0.00	0.00	0.01	0.02	0.06	0.13	0.21	0.25	0.22	0.10
0.00	0.00	0.00	0.01	0.02	0.05	0.11	0.19	0.26	0.24	0.13
0.00	0.00	0.00	0.00	0.01	0.04	0.10	0.18	0.26	0.26	0.15
0.00	0.00	0.00	0.00	0.01	0.03	0.09	0.17	0.26	0.27	0.16
0.00	0.00	0.00	0.00	0.01	0.03	0.08	0.16	0.26	0.28	0.18
0.00	0.00	0.00	0.00	0.01	0.03	0.08	0.16	0.25	0.29	0.18
0.00	0.00	0.00	0.00	0.01	0.03	0.08	0.16	0.25	0.29	0.18
0.00	0.00	0.00	0.00	0.01	0.03	0.08	0.16	0.26	0.28	0.18
0.00	0.00	0.00	0.00	0.01	0.04	0.09	0.17	0.26	0.27	0.16
0.00	0.00	0.00	0.00	0.01	0.04	0.10	0.18	0.26	0.26	0.14
0.00	0.00	0.00	0.01	0.02	0.05	0.11	0.19	0.25	0.24	0.12
0.00	0.00	0.00	0.01	0.03	0.07	0.13	0.21	0.25	0.21	0.10
0.00	0.00	0.00	0.01	0.04	0.08	0.15	0.22	0.24	0.18	0.08
0.00	0.00	0.01	0.02	0.05	0.11	0.17	0.22	0.22	0.15	0.06
0.00	0.00	0.01	0.03	0.07	0.13	0.19	0.22	0.19	0.12	0.04
0.00	0.00	0.02	0.05	0.09	0.15	0.20	0.21	0.16	0.09	0.03
0.00	0.01	0.03	0.06	0.12	0.17	0.21	0.19	0.13	0.06	0.02
0.00	0.01	0.04	0.08	0.14	0.19	0.20	0.17	0.11	0.05	0.01
0.00	0.02	0.05	0.11	0.16	0.20	0.19	0.14	0.08	0.03	0.01
0.01	0.03	0.07	0.13	0.18	0.20	0.18	0.12	0.06	0.02	0.00
0.01	0.03	0.08	0.14	0.19	0.20	0.16	0.11	0.05	0.02	0.00
0.01	0.04	0.10	0.16	0.20	0.20	0.15	0.09	0.04	0.01	0.00
0.01	0.05	0.11	0.17	0.20	0.19	0.14	0.08	0.04	0.01	0.00
0.01	0.05	0.11	0.17	0.20	0.19	0.14	0.08	0.03	0.01	0.00
0.01	0.05	0.11	0.17	0.21	0.19	0.14	0.08	0.03	0.01	0.00
0.01	0.05	0.11	0.17	0.20	0.19	0.14	0.08	0.04	0.01	0.00
0.01	0.04	0.10	0.16	0.20	0.20	0.15	0.09	0.04	0.01	0.00
0.01	0.03	0.09	0.15	0.20	0.20	0.16	0.10	0.05	0.02	0.00
0.01	0.03	0.07	0.13	0.18	0.20	0.17	0.12	0.06	0.02	0.00
0.00	0.02	0.06	0.11	0.17	0.20	0.19	0.14	0.08	0.03	0.01
0.00	0.01	0.04	0.09	0.15	0.19	0.20	0.16	0.10	0.04	0.01
0.00	0.01	0.03	0.07	0.13	0.18	0.21	0.18	0.12	0.06	0.01
0.00	0.00	0.02	0.05	0.10	0.16	0.20	0.20	0.15	0.08	0.02
0.00	0.00	0.01	0.03	0.08	0.14	0.19	0.22	0.18	0.11	0.03

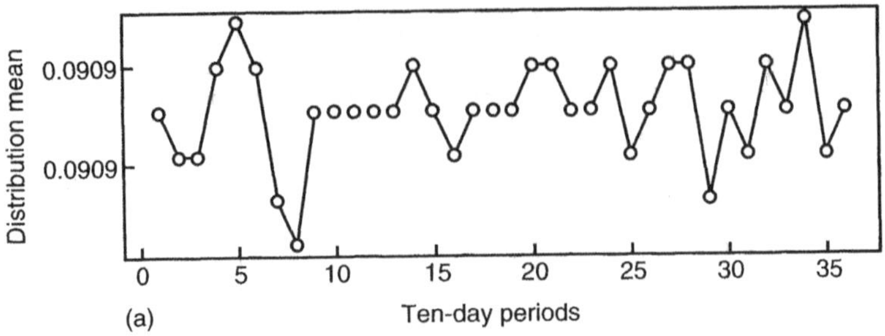

Figure 3.5a Means of the 36 distributions of numbers of wet days in each of 36 consecutive 10-day periods: Manaus daily rainfall data

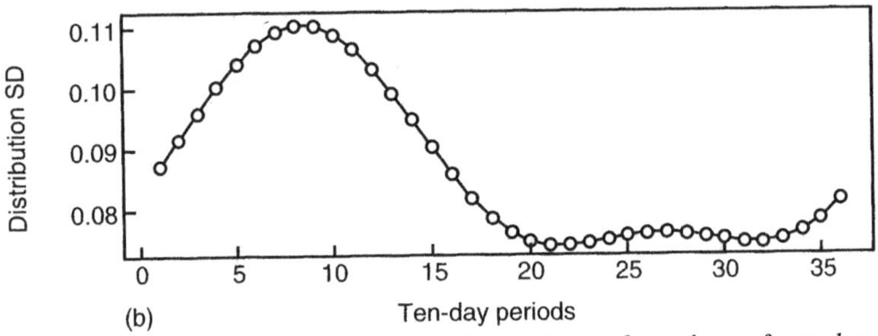

Figure 3.5b Standard deviations of the 36 distributions of numbers of wet days in each of 36 consecutive 10-day periods: Manaus daily rainfall data

All other q_{ijkl} are zero, and the only other terms necessary to calculate the recurrence relation are the transition probabilities $p_{00}(t)$, $p_{01}(t)$, $p_{10}(t)$ and $p_{11}(t)$, and the initial state. Considering again the table of initial probabilities:

$P[Y=1, J=0]$	$P[Y=2, J=0]$	$\ldots$	$P[Y=K, J=0]$	$P[J=0]$
$P[Y=1, J=1]$	$P[Y=2, J=1]$	$\ldots$	$P[Y=K, J=1]$	$P[J=1]$
$P[Y=1]$	$P[Y=2]$	$\ldots$	$P[Y=K]$	1

(3.36)

where $Y \equiv Y(t_0)$, $J \equiv J(t_0)$

We can start the calculation conditional on day t_0 being (say) wet by setting $P[Y(t_0)=1, J(t_0)=1]$ in this table; all other probabilities will be zero within the body of the table, and in the margins only $P[Y(t_0)=1]=1$ and $P[J(t_0)=1]=1$. The recurrence relation then recomputes this table at times

$t_0 + 1, t_0 + 2, \ldots t_0 + m$. The quantity that we require, at the end of the m repetitions, is $P[Y(t_m) = K]$, and this is obtained from the bottom margin of the table.

As an example, we consider the models (3.31a) and (3.31b) representing the occurrence of rainfall at Reserva Ducke, Manaus. We take as the model for $p_{01}(t)$ and $p_{11}(t)$ the following:

$$\log_e[p_{01}(t)/(1 - p_{01}(t))] = 0.1551 + 0.1584\cos(2\pi t/365) + 0.7292\sin(2\pi t/365)$$

and

$$\log_e[p_{11}(t)/(1 - p_{11}(t))] = 0.9976 + 0.1612\cos(2\pi t/365) + 0.8050\sin(2\pi t/365)$$

and take $K = 5$, $m = 15$; thus we calculate the probability of five or more consecutive dry days, in a succession of 15-day periods throughout the year, assuming that day t_0 is wet in each case. The calculated probabilities are shown in Figure 3.6.

3.8.3 Example: calculating the probability distribution of soil water content

The examples in sections 3.8.1 and 3.8.2 were simple in that they used only the model for rainfall occurrence: the model fitted to rainfall amounts did not enter the recursive calculation. Stern and Coe (1984) present a third example of the use of the recurrence relation (3.24) in which both models – for rainfall occurrence and for rainfall depth – are used to calculate the probability distribution of the depth of soil water stored within a soil layer. Water in this layer is increased by rainfall, and decreased by evaporation; the model is appropriate for the semi-arid conditions for which the authors developed it, but is oversimplified for modelling soil-water behaviour in wetter climates where, in addition to evaporation loss, water is also lost by surface runoff when soil is saturated, and by percolation to deeper soil layers. Adaptation of the basic Stern–Coe model to deal with the first of these phenomena – surface runoff from saturated soil – is discussed in a later chapter, where some analytical results are obtained for simple cases. More realistic models of soil-water behaviour, and its relation to river runoff and evaporation loss, must make extensive use of simulation, both to estimate the parameters defining the relationships between storage and outflow (for example, the parameters describing how evaporation is reduced when soil moisture is limiting), and to estimate the frequencies with which events of hydrological importance occur.

Denoting the depth of water in the soil layer by $W(t)$, we have the water balance equation

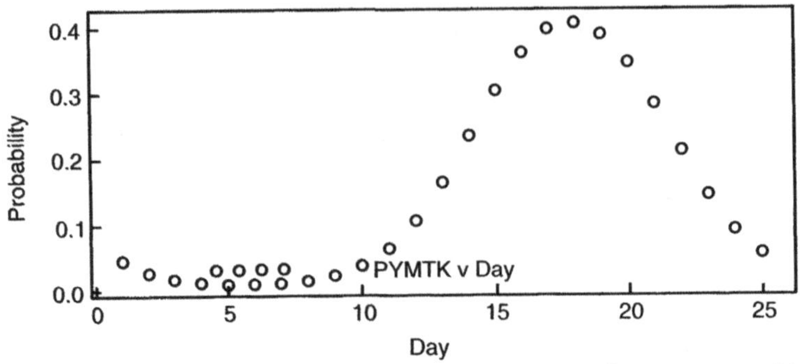

Figure 3.6 Probability of five or more consecutive dry days in a sequence of 15-day periods throughout the year: Reserva Ducke, Manaus, Amazonas

$$W(t) = W(t-1) + X(t) - e, \tag{3.37}$$

where rainfall $X(t)$ is zero when $J(t) = 0$. Soil water $W(t)$ is constrained to be not less than zero, and not greater than some maximum value W_{max} (analogous to "field capacity"). The quantity e in (3.37) is the loss by evaporation (and by other hydrological action such as deep percolation and runoff); in the example discussed by Stern and Coe (1984), e is assumed constant, but at the cost of slightly greater complexity it can be replaced by a known time-varying function $e(t)$, which may allow for differences in evaporation loss on wet days and dry days. If the recurrence relation (3.24) is to be used, it does not appear to be possible to let $e(t)$ depend on $W(t)$.

In the recurrence relation (3.24) the random variable $Y(t)$ is now taken to have $K + 1$ discrete states representing the depth of water stored in the soil profile; the state $Y(t) = 0$ represents absence of water in the profile, $Y(t) = K = W_{max}$ corresponds to saturation. It is important that the discrete intervals in the range 0 to W_{max}, and the loss e, be measured in the same units; millimetres is the obvious choice.

The calculation uses the numerical values of upper-tail probabilities defined by

$$G_{0y}(t) = P[Y(t) \geq y - \tfrac{1}{2} \mid J(t-1) = 0] \tag{3.38a}$$

and

$$G_{1y}(t) = P[Y(t) \geq y - \tfrac{1}{2} \mid J(t-1) = 1] \tag{3.38b}$$

These are derived from the gamma distributions fitted to the rainfall depths on days following a dry day (in the case of (3.38a)) and a wet day ((3.38b)) respectively, using the model fitted to mean rainfall $\mu(t)$, and the estimated shape coefficient κ. The probabilities $G_{0y}(t)$ and $G_{1y}(t)$ are used to define the quantities

$$q_{ijkl} = P[Y(t) = l \mid Y(t-1) = k, J(t) = j, J(t-1) = i]$$

in (3.27). With the evaporation loss e assumed constant, and $Y(t-1)$ in state k, we have (omitting the ever-present argument t)

$$q_{00k(k-e)} = 1 \quad \text{and} \quad q_{10k(k-e)} = 1 \quad \text{when } k > e \tag{3.39}$$

expressing the fact that soil water is reduced by the loss e when there is no rain on day t, $Y(t)$ thereby being in state $k - e$ when this is positive. When the evaporative demand is greater than the soil-water content k, however (i.e. when $k \leq e$), Y(t) must move to its zeroth state with probability one; that is,

$$q_{00k0} = 1 \quad \text{and} \quad q_{10k0} = 1. \tag{3.40}$$

The probabilities (3.39) and (3.40) all refer to what happens when day t is dry. When rain falls on day t, soil water increases; provided that the amount of rain is more than sufficient to satisfy evaporation demand and less than that required to fill the soil layer to saturation, $Y(t-1)$ then moves from state k to some higher state, r say, such that

$$q_{01k(k+r-e)} = G_{0(r+1)} - G_{0r}$$

$$\text{(provided } k + r - e > 0 \text{ and } k + r - e < K) \tag{3.41}$$

when day $t - 1$ was dry, whilst when day $t - 1$ was wet we have the equivalent

$$q_{11k(k+r-e)} = G_{1(r+1)} - G_{1r}$$

$$\text{(provided } k + r - e > 0 \text{ and } k + r - e < K) \tag{3.42}$$

with the $G(.)$ functions defined by (3.38a,b). When, however, the rain on day t is insufficient to satisfy the evaporative demand on that day, soil water falls from state k to zero, despite the small rainfall, with probability that the rain that falls is less than $e - k$ (the excess of evaporative demand over soil moisture content on day $t - 1$):

$$q_{01k0} = 1 - G_{0(e-k)} \quad \text{and} \quad q_{11k0} = 1 - G_{1(e-k)} \quad \text{when } k \leq e \tag{3.43}$$

Finally, the case where rain on day t is so large that the soil layer is filled to its maximum possible value W_{max}: the soil moisture store fills to capacity with probability equal to the probability that the rainfall input is greater than the available empty storage $K - k - e$, allowing for evaporative loss:

$$q_{11kK} = 1 - G_{1(K-k-e)} \quad \text{and} \quad q_{01kK} = G_{0(K-k-e)}. \tag{3.44}$$

The remaining q_{ijkl} are zero. Knowledge of the distributions required for (3.38a,b) and of the non-zero q_{ijkl} given in (3.39) to (3.44), together with the initial soil moisture condition, allow us to use the recurrence relation (3.24) to calculate the probabilities $P[Y(t) \leq y, J(t) = 0]$ for the distribution function of

$Y(t)$ when day t is dry, and $P[Y(t) \leq y, J(t) = 1]$, the distribution function of soil water content when day t is wet.

An important aspect of the recursive calculation is the choice of the value of K, the number of states of the random variable $Y(t)$; this controls the substitution of the continuous distribution of rainfall depth by a discrete distribution. In practice, rainfall is effectively measured at discrete intervals in any case. For semi-arid environments, Stern and Coe report good results for $K = 100$ mm. As noted above, the recursive method requires only minor modification to deal with cases where evaporation loss e is a time-variant function, possibly different for wet and dry days; a slight modification also allows e to be allowed to vary randomly. Limitations of the recursive method in wetter climates, however, are likely to be the difficulty of relating losses e to the soil-water content $Y(t)$, and the difficulty of dealing with losses other than through evaporation.

3.9 MODIFICATION OF STERN–COE MODELS TO EXAMINE TIME TRENDS IN RAINFALL OCCURRENCE AND AMOUNT

McCullagh and Nelder (1989) point out that it is straightforward to modify the Stern–Coe models described in this chapter to examine whether time trends exist in rainfall occurrence and amount over a period of (say) N years of record. Take rainfall occurrence first, and assume a first-order Markov chain, and no missing data, for simplicity: in estimating the transition probabilities, the Stern–Coe models described above consider the N transitions from 1 January to 2 January, from 2 January to 3 January, . . . as N replicates of the same event, which are used to estimate the transition probabilities $p_{00}, p_{01}, p_{10}, p_{11}$. Where a time trend is present, however, the N transitions are no longer replicates, and the presence and absence of rain in the period of record must be considered as $365N$ observations of a (dependent) Bernoulli process, modelled (with the assumptions noted above) as

$$\begin{aligned} \log_e[p_{01}(t)/\{1 - p_{01}(t)\}] &= y_s + \gamma^T \mathbf{x} \\ \log_e[p_{11}(t)/\{1 - p_{11}(t)\}] &= y_s + \phi^T \mathbf{x}, \qquad s = 1, 2, \ldots N, \end{aligned} \tag{3.45}$$

where y_s is a "year effect"; in other words, the 0–1 variables in the sequence of rainfall occurrences is indexed now by year, day within year (recalling that $\mathbf{x}$ is a vector of harmonic terms) and previous day wet or dry. A statistical test for the absence of year effects is achieved by testing the null hypothesis H_0: $y_1 = y_2 = \ldots y_N = 0$ against the alternative H_1: $y_1 \neq y_2 \neq \ldots y_N \neq 0$. If the null hypothesis is rejected, further tests will be required to ascertain whether the rainfall trend is linear or of some more complicated polynomial shape. Similarly, the model for the quantity of rainfall

$$\log_e[\mu_{01}(t)] = \boldsymbol{\lambda}^T\boldsymbol{x} \tag{3.46}$$

(where $\boldsymbol{\lambda}$ is a vector of coefficients and $\boldsymbol{x}$ is a vector of harmonic terms) is replaced by

$$\log_e[\mu_{01}(s,t)] = y_s + \boldsymbol{\lambda}^T\boldsymbol{x} \qquad s = 1, 2, \ldots N. \tag{3.47}$$

Once again, the test of H_0: $y_1 = y_2 = \ldots y_N = 0$ against the alternative H_1: $y_1 \neq y_2 \neq \ldots y_N \neq 0$ may be used to test whether there are significant year effects in depth of rain falling after a dry day, and after a preceding wet day; if the hypothesis is rejected, further tests will determine the form of the trend. In each case, the test is made by calculating the deviance statistic under the two hypotheses H_0 and H_1, the increase in deviance associated with H_0 being approximately distributed as χ^2 with degrees of freedom equal to the difference between the degrees of freedom of the deviance under H_0 minus those for H_1.

3.10 COMPUTER METHODS FOR GENERATION OF ARTIFICIAL RAINFALL SEQUENCES

The generation of artificial sequences of daily rainfall by use of Stern–Coe models requires (1) the generation of sequences for the occurrence of rainfall, and (2) generation of rainfall depths on wet days. We deal briefly with the two steps in the following.

1. Generation of sequences of rainfall occurrence

For a two-state, first-order, time-variant Markov chain, recall that the 2×2 matrix of transition probabilities has the form

$$\mathbf{P}(t) = \text{State at time } t-1 \begin{matrix} \\ 0 \\ 1 \end{matrix} \overset{\text{State at time } t}{\begin{matrix} \quad 0 \qquad\quad 1 \\ \begin{bmatrix} p_{00}(t) & p_{01}(t) \\ p_{10}(t) & p_{11}(t) \end{bmatrix} \end{matrix}}$$

If the fitted model is, say, of the form

$$\log_e[p_{01}(t)/\{1 - p_{01}(t)\}] = \beta_0 + \beta_1\cos(2\pi t/365) + \beta_2\sin(2\pi t/365) \tag{3.48a}$$

$$\log_e[p_{11}(t)/\{1 - p_{11}(t)\}] = \gamma_0 + \gamma_1\cos(2\pi t/365) + \gamma_2\sin(2\pi t/365) \tag{3.48b}$$

then $p_{01}(t), p_{11}(t)$ are given by

$$p_{01}(t) = \exp(B)/[1 + \exp(B)]$$
$$p_{11}(t) = \exp(G)/[1 + \exp(G)],$$

where B, G are the right-hand sides of (3.48a,b).

If day $t - 1$ was dry (state 0), the state on day t is given by generating a uniformly distributed pseudo-random variable on the interval (0, 1); if the generated value is less than or equal to $p_{00}(t)$, day t is taken to be dry (state 0), else it is taken as wet (state 1). If day $t - 1$ was wet (state 1), day t is taken as dry (state 0) if the generated value is less than $p_{10}(t)$, else day t is taken as wet (state 1).

The generated state on day t is then used to enter the 2×2 matrix of transition probabilities for transitions between the two states on day t and those on day $t + 1$, the process being repeated until a sequence of the desired length has been generated.

2. *Generation of rainfall depths on wet days*

If the generation of the 0–1 sequence of dry and wet days gives a wet day on day t, a value from the fitted gamma distribution must be generated with the appropriate parameters $\mu(t)$ and κ. If the $\mu(t)$ has been modelled as in (3.17), it will be given by $\exp(H)$, where, from (3.17)

$$H = \lambda_0 + \lambda_1 \cos(2\pi t/365) + \lambda_2 \sin(2\pi t/365) \\ + \lambda_3 \cos(4\pi t/365) + \lambda^4 \sin(4\pi t/365) + \dots$$

The parameter κ, if assumed constant throughout the year as in section 3.7, then completes the necessary specification of the gamma distribution, from which the pseudo-random variable can be obtained.

3.11 CONCLUSION

This chapter has given a fairly detailed account of the use of Markov chains for the modelling of daily rainfall, and for the use of such models to answer questions that are of great importance to agricultural planners, using methods propounded by Stern and Coe (1984). Their methods were developed from earlier studies (Chin, 1977; Todorovic and Woolhiser, 1975; Woolhiser and Pegram, 1979) where Markov chains were used to model daily rainfall, and where in many cases the exponential distribution was assumed for rainfall depths; however, the use of the gamma distribution (Ison et al., 1971; Buishand, 1977) is usually recommended instead of the exponential – one of the strengths of the Stern–Coe approach – and this was confirmed in the example worked in this chapter of daily rainfall at Reserva Ducke, Manaus.

Regarding rainfall occurrence, we have assumed throughout that the Markov chain had just two states (0, 1). In practice, as Stern and Coe demonstrate, it may be necessary to use three states (0, "trace", 1) to allow for the greater errors when measuring very small amounts of rainfall. In this context, the work of Gregory et al. (1993) must be mentioned; these authors concluded that instead of a two-state process requiring a second- or higher-order chain, it might be preferable to use a multiple-state process of first order.

Chapter Four

More on Markov chains: storage

4.1 MARKOV CHAINS WITH r STATES

Chapter Three dealt mainly with the simple case of Markov chains with two states, typically the occurrence or non-occurrence of rain, for which the state random variable assumed values 0 or 1. The generalisation to r states ($r > 2$) is straightforward: the states of the chain are now the values $0, 1, \ldots r - 1$ taken by a random variable X; the matrix of transition probabilities $\mathbf{P}$ typically has the element p_{ij} in the ith row and jth column (the sum of the elements in each row being one), representing the probability of a transition from state i at time $t - 1$, to state j at time t.

4.2 DEFINITIONS

Before quoting briefly and without proof some results from the theory of homogeneous (time-invariant) r-state Markov chains, we require some definitions. Suppose that the chain is initially in state j at time $t = 0$, state j being one of the r states associated with the random variable X taking possible values $0, 1, 2, \ldots r - 1$. If a return to state j at some future time is a certain event, then the state j is a **recurrent state**; the time taken to return to state j is then a random variable called the **recurrence time**. The state j is termed **positive-recurrent** if the mean recurrence time is finite, else it is **null-recurrent**. If a return to state j has probability less than unity, the state is called **transient**. The state j is termed **ephemeral** if, for all other states i, $p_{ij} = 0$, signifying that the state j can never be reached from any other state. (The chain may be in state j at time $t = 0$, passing out of it upon the first transition; but if the state is ephemeral, the chain will never subsequently return to it.)

Remember that the chain is assumed homogeneous, i.e. transition probabilities are time-invariant. With the state in state j at time $t = 0$, let $f_{jj}^{(n)}$ be the probability that the next occupancy of state j is at time $t = n$. Clearly for

$n = 1$, we must have $f_{jj}^{(1)} = p_{jj}$, the probability of a transition from state j at time $t = 0$ to the same state at time $t = 1$; but for $n > 1$

$$f_{jj}^{(n)} = P[(X_1 \neq j) \cap (X_2 \neq j) \ldots \cap (X_{n-1} \neq j) \cap (X_n = j) | (X_0 = j)]$$

We see that $f_j = \Sigma f_{jj}^{(n)}$, summed over n, is the probability that the state j is eventually re-entered: if f_j has value one, state j is recurrent, otherwise state j is transient. When the state j is recurrent ($f_j = 1$), so that its future occurrence is a certain event, $f_{jj}^{(n)}$ has a probability distribution for each of the values $n = 1, n = 2, \ldots$; for any one value of n, therefore, the mean value of this probability distribution is

$$\mu_j = \sum_{n=1}^{\infty} n f_{jj}^{(n)}.$$

This is the **mean recurrence time**, the mean time taken to return to state j following an excursion from it. We can generalise the above by defining

$$f_{jk}^{(n)} = P[(X_1 \neq k) \cap (X_2 \neq k) \ldots \cap (X_{n-1} \neq k) \cap (X_n = k) | (X_0 = j)]$$

to be the probability of entering state k for the first time, given that the initial state is j. Then

$$f_{jk} = \sum_{n=1}^{\infty} f_{jk}^{(n)}$$

is the probability that state k is ever entered; if $f_{jk} = 1$, entry to state k at some future time, given the chain in state j initially, is certain, and the random variable $f_{jk}^{(n)}$ has a probability distribution with mean value

$$\mu_{jk} = \sum_{n=1}^{\infty} n f_{jk}^{(n)}.$$

This is the **mean first passage time**, the mean time to the first entry of state k, given that the chain is initially in state j.

Finally, denote by $F_{jj}(s)$ the probability generating function (see Chapter Two) of the probabilities $f_{jj}^{(n)}$, so that

$$F_{jj}(s) = \sum_{n=1}^{\infty} f_{jj}^{(n)} s^n. \tag{4.1}$$

Similarly, define

$$F_{jk}(s) = \sum_{n=1}^{\infty} f_{jk}^{(n)} s^n \tag{4.2}$$

$$P_{jj}(s) = \sum_{n=1}^{\infty} p_{jj}^{(n)} s^n \tag{4.3}$$

$$P_{jk}(s) = \sum_{n=1}^{\infty} p_{jk}^{(n)} s^n. \tag{4.4}$$

The probability generating functions (4.3) and (4.4) generate the nth powers of the transition probabilities p_{jj} and p_{jk}, the elements of the transition probability matrix **P**. Specifically, the ***n*-step transition probabilities**

$$p_{jk}^{(n)} = P\,[\text{state } k \text{ at time } n \mid \text{state } j \text{ at time } 0\,].$$

are the elements of the matrix $\mathbf{P}^n$, the nth power of **P**.

4.2.1 Example

Example 1: Rodriguez-Iturbé and Valdes (1979) proposed a probabilistic model of river basin behaviour, in which the state variable X is the order of the stream in which a water particle is located at time t. For a basin of order Ω, there are $N = \Omega + 1$ states; a particle may originate in a sub-basin of any order, but all particles eventually terminate their transitions (are **absorbed**) on reaching state N. Thus the matrix of transition probabilities has the form

$$\mathbf{P} = \begin{bmatrix} p_{11} & p_{12} & p_{13}\cdots & p_{1N} \\ p_{21} & p_{22} & p_{23}\cdots & p_{2N} \\ \cdots & & \cdots & \\ P_{N1} & P_{N2} & P_{N3}\cdots & P_{NN} \end{bmatrix} = \begin{bmatrix} \Theta & p_{12} & p_{13}\cdots & p_{1\Omega} & 0 \\ 0 & 0 & p_{23}\cdots & p_{2\Omega} & 0 \\ \cdots & & \cdots & & \\ 0 & 0 & 0\ldots & 0 & 1 \end{bmatrix}$$

It is clear that state 1 is an ephemeral state, since it cannot be entered from states $2, 3, \ldots N$; none of the states is recurrent, since once a state is entered it can never be re-entered. All of the $f_{jj}^{(n)}$ are zero, so that the probability generating function $F_{jj}(s)$ is also zero, although the probability generating functions $F_{jk}(s)$ are not zero, in general, for j and k unequal.

4.3 SOME STANDARD RESULTS FOR HOMOGENEOUS (TIME-INVARIANT) *r*-STATE MARKOV CHAINS

Theory (see, for example, Cox and Miller, 1965; Grimmett and Stirzaker, 1992) proves the following results, for the case where probabilities of transitions between states are constant and not time-variant:

1. If the row vector $\boldsymbol{p}^{(n)} = [p_0^{(n)}, p_1^{(n)}, \ldots p_r^{(n)}]$ denotes the probabilities of the system being in states $0, 1, 2 \ldots r$ at time n, with initial states given in the row vector

$$\boldsymbol{p}^{(0)} = [p_0^{(0)}, p_1^{(0)} \ldots p_r^{(0)}],$$

then

$$\boldsymbol{p}^{(n)} = \boldsymbol{p}^{(n-1)}\mathbf{P}, \tag{4.5}$$

where **P** is the matrix of transition probabilities given by $\mathbf{P} = [p_{ij}]$. Hence, by repeated substitution,

$$p^{(n)} = p^{(n-1)}\mathbf{P} = p^{(n-2)}\mathbf{P}^2 = \ldots = p^{(0)}\mathbf{P}^n. \tag{4.6}$$

This simply restates a result quoted in Chapter 3.

2. Assuming that the matrix **P** has distinct eigenvalues, a general way of calculating the matrix $\mathbf{P}^n$ of n-step transition probabilities is given by

$$\mathbf{P} = \mathbf{Q}\mathbf{\Lambda}\mathbf{Q}^{-1},$$

where $\mathbf{\Lambda}$ is the matrix of eigenvalues $\mathbf{\Lambda} = \text{diag}(\lambda_1, \lambda_2, \ldots \lambda_r)$ and **Q** is the matrix having the r distinct eigenvectors as its columns, $\mathbf{Q} = [e_1, e_2, \ldots e_r]$.

3. Since $P^{m+n} = \mathbf{P}^m\mathbf{P}^n$, the $(m+n)$-step transition probabilities can be expressed in terms of the m-step and n-step transition probabilities by equating terms on either side of this equation, so that

$$p_{jk}^{(m+n)} = \sum_l p_{jl}^{(m)} p_{lk}^{(n)} \tag{4.7}$$

the **Chapman–Kolmogorov relation** which all Markov chains satisfy.

4. The probability generating functions $F_{jj}(s)$ and $P_{jj}(s)$, and the probability generating functions $F_{jk}(s)$ and $P_{jk}(s)$, are related by

$$F_{jj}(s) = P_{jj}(s)/[1 + P_{jj}(s)] \tag{4.8}$$

and

$$F_{jk}(s) = P_{jk}(s)/[1 + P_{jk}(s)]. \tag{4.9}$$

Therefore, by expanding the right-hand sides of (4.8) and (4.9) in powers of s and equating coefficients of s^n, explicit values can be obtained for $f_{jj}^{(n)}$ and $f_{jk}^{(n)}$, in terms of (powers of) the known transition probabilities p_{jj} and p_{jk}.

4.3.1 Example

We illustrate the use of (4.8) and (4.9) by applying them to the simple case of a chain with two states only, as considered in Chapter Three. The 2×2 matrix of transition probabilities given by Gabriel and Neumann (1962) for the sequence of wet and dry days at Tel Aviv is

$$\mathbf{P} = \begin{bmatrix} 0.750 & 0.250 \\ 0.338 & 0.662 \end{bmatrix}$$

so that the probability of a run of three wet days, given that the weather is dry on day $t = 0$, is clearly

$$0.25 \times 0.662 \times 0.662 \times 0.338 = 0.037\,032.$$

To calculate the probabilities of runs of wet days, given a chain in state 0 at time $t = 0$, we need to calculate terms in the expansion of

$$F_{00}(s) = P_{00}(s)/[1 + P_{00}(s)]$$

where

$$P_{00}(s) = \sum_{n=1}^{\infty} p_{00}^{(n)} s^n$$

and $p_{00}^{(n)}$ are the elements in the first row, first column of nth powers of the matrix **P**. Calculating these powers, we have

$$\mathbf{P}^2 = \begin{bmatrix} 0.647000 & 0.353000 \\ 0.477256 & 0.522744 \end{bmatrix} \quad \mathbf{P}^3 = \begin{bmatrix} 0.604564 & 0.395436 \\ 0.534629 & 0.465371 \end{bmatrix}$$

$$\mathbf{P}^4 = \begin{bmatrix} 0.587080 & 0.412920 \\ 0.558267 & 0.441733 \end{bmatrix} \quad \mathbf{P}^5 = \begin{bmatrix} 0.579877 & 0.420123 \\ 0.568006 & 0.431994 \end{bmatrix}$$

and so on. Hence, by the definition of $P_{00}(s)$, we have

$$P_{00}(s) = 0.750s + 0.647000s^2 + 0.604564s^3 + 0.587080s^4 + 0.579877s^5 + \ldots$$

and using

$$F_{00}(s) = P_{00}(s)/[1 + P_{00}(s)]$$

we have the identity

$$(f_{00}^{(1)}s + f_{00}^{(2)}s^2 + f_{00}^{(3)}s^3 + \ldots)(1 + 0.750\,s + 0.647000s^2 + 0.604564s^3 + 0.587080s^4 + 0.579877s^5 + \ldots$$

$$= 0.750s + 0.647000s^2 + 0.604564s^3 + 0.587080s^4 + 0.579877s^5 + \ldots$$

whence, equating coefficients of powers of s on each side, we have

$$f_{00}^{(1)} = 0.75$$
$$f_{00}^{(2)} + 0.75f_{00}^{(1)} = 0.647$$
$$f_{00}^{(3)} + 0.75f_{00}^{(2)} + 0.647f_{00}^{(1)} = 0.604564$$
$$f_{00}^{(4)} + 0.75f_{00}^{(3)} + 0.647f_{00}^{(2)} + 0.604564f_{00}^{(1)} = 0.587080$$

and so on. Solving by repeated substitution, we have

$$f_{00}^{(1)} = 0.75; f_{00}^{(2)} = 0.0845;$$

$$f_{00}^{(3)} = 0.05594; f_{00}^{(4)} = 0.037030; \ldots.$$

Hence we see that $f_{00}^{(4)}$ agrees with the value obtained previously by simple calculation.

Similarly, using the definitions

$$P_{11}(s) = 0.662s + 0.522744s^2 + 0.465371s^3 + 0.441733s^4 + 0.431733s^5 + \ldots$$
$$P_{01}(s) = 0.250s + 0.353000s^2 + 0.395436s^3 + 0.412920s^4 + 0.420123s^5 + \ldots$$
$$P_{10}(s) = 0.338s + 0.477256s^2 + 0.534629s^3 + 0.558267s^4 + 0.568006s^5 + \ldots$$

we can calculate the values of $f_{11}^{(n)}$, the probabilities of first return to state 1 at time n given that the chain is initially in state 1, and $f_{01}^{(n)}$ and $f_{10}^{(n)}$, the probabilities that state 1 (0) is first entered at time n, given that the chain is initially in state 0 (1).

4.4 NON-HOMOGENEOUS *r*-STATE MARKOV CHAINS

The material so far presented in this chapter is a short summary of some of the more common results applicable to homogeneous Markov chains. In many practical applications, homogeneity in time cannot be assumed, and the relation $\boldsymbol{p}^{(n)} = \boldsymbol{p}^{(0)}\mathbf{P}^n$ given in (4.6) does not hold. If, however, we write $p_{ij}(t+1)$ for the transition probability from state i to state j in the time interval $(t, t+1)$, generalisation of (4.6) is straightforward, at least in theory. If $\boldsymbol{p}^{(0)} = [p_0^{(0)}, p_1^{(0)} \ldots p_r^{(0)}]$ is the row vector of probabilities of the chain being in states $0, 1, 2 \ldots r$ at time $t = 0$, and denoting by $\mathbf{P}(t, t+1)$ the matrix of transition probabilities with (i, j)th element $p_{ij}(t, t+1)$, we must have

$$\boldsymbol{p}^{(t)} = \boldsymbol{p}^{(t-1)}\mathbf{P}(t-1, t)$$

with

$$\boldsymbol{p}^{(2)} = \boldsymbol{p}^{(1)}\mathbf{P}(1, 2) \qquad (t = 1)$$
$$\boldsymbol{p}^{(1)} = [p_0^{(0)}, p_1^{(0)} \cdots p_0^{(0)}].\mathbf{P}(0, 1) \qquad (t = 0)$$

Hence we have

$$\boldsymbol{p}^{(t)} = [p_0^{(0)}, p_1^{(0)} \ldots p_0^{(0)}].\mathbf{P}(0, 1).\mathbf{P}(1, 2).\ldots.\mathbf{P}(t-2, t-1)\mathbf{P}(t-1, t) \quad (4.10)$$

which is a straightforward generalisation of (4.6). In practice, the estimates of the matrices $\mathbf{P}(t-1, t)$ of transition probabilities may be poor, particularly if the number of states r is large and the transitions between some of them very infrequent. Suppose, by way of illustration, that daily rainfall is divided into 10 states, the first corresponding to no rain, the second to a "trace", say less than 2 mm, and the tenth state corresponding to a rainfall greater than 50 mm (states 3 to 9 being some appropriate subdivision of the interval between 2 and 50 mm). Proceeding as shown in Chapter Three, it would be possible to use data from 1 and 2 January in the existing rainfall record, to estimate the transition probabilities between the 10 states for these two days, thus obtaining $\mathbf{P}(0, 1)$; data from 2 and 3 January would yield estimates of transition probabilities in the matrix $\mathbf{P}(1, 2)$ and so on. Whilst in Chapter Three we were estimating only two transition probabilities $p_{01}(t, t+1)$ and $p_{11}(t, t+1)$ – since

$$p_{00}(t, t+1) = 1 - p_{01}(t, t+1); p_{10}(t, t+1) = 1 - p_{11}(t, t+1)$$

– we are now estimating 10×9 transition probabilities. If, say, 20 years of rainfall record are available, so that there are 20 observed transitions from the state on 1 January to 2 January, clearly many of the elements in the matrix $\mathbf{P}(0, 1)$ will remain at zero. With many states, therefore, very long records are necessary to obtain estimates of the transition probabilities of a non-homogeneous Markov chain; in practice it will often be necessary to assume "piece-wise homogeneity": for example, that the chain is homogeneous during three-month periods of the year. It is also worth pointing out that in (4.10) it is not essential that all the transition matrices $\mathbf{P}(t, t+1)$ be square, provided that the product is conformable for matrix multiplication (i.e. $\mathbf{P}(t, t+1)$ must always have the same number of rows as $\mathbf{P}(t-1, t)$ has columns, and the same number of columns as $\mathbf{P}(t+1, t+2)$ has rows).

When modelling the occurrence of rainfall between October and May at Irbid in Jordan, Stern and Coe (1984) found that a first-order chain was adequate if three states were used: 0 ("dry"), 1 ("trace", denoting rainfall less than 2.5 mm, including entries marked as "trace" in the field record), and 2 ("wet"). The transition matrix between days t and $t+1$ was then the 3×3 matrix

$$\mathbf{P}^2 = \begin{bmatrix} p_{00}(t, t+1) & p_{01}(t, t+1) & p_{02}(t, t+1) \\ p_{10}(t, t+1) & p_{11}(t, t+1) & p_{12}(t, t+1) \\ p_{20}(t, t+1) & p_{21}(t, t+1) & p_{22}(t, t+1) \end{bmatrix}$$

Altogether six transition probabilities need to be modelled: namely, $p_{01}(t, t+1)$, $p_{02}(t, t+1)$, $p_{11}(t, t+1)$, $p_{12}(t, t+1)$, $p_{21}(t, t+1)$ and $p_{22}(t, t+1)$ (since the probabilities in each row add to one). Seasonal variation in these time-dependent probabilities was modelled by logistic regression on the harmonics

$$\cos(2\pi t/365), \sin(2\pi t/365), \ldots$$

By comparing the fitted curves, Stern and Coe found that the six curves could be reduced to five: three for modelling

$$p_{02}(t, t+1), p_{12}(t, t+1) \quad \text{and} \quad p_{22}(t, t+1),$$

one for modelling $p_{01}(t, t+1)$, and one for modelling

$$p_{11}(t, t+1) \quad \text{and} \quad p_{21}(t, t+1).$$

The method for comparing the fitted curves is as illustrated in Chapter Three. Dummy variables are introduced, for the purpose of testing whether the fitted harmonic curves are parallel; if the test for parallelism shows no significant increase in the deviance statistic, the curves are taken to be parallel and the test for displacement between the parallel curves follows. If the test for displacement shows no significant increase in the deviance statistic, the parallel curves collapse into a single curve.

4.5 ASPECTS OF THE THEORY OF STORAGE: THE MORAN ANNUAL MODEL

For almost half a century, Markov chains have been extensively used to describe the time variation of water volumes stored in reservoirs that are supplied by river inflows subject to random fluctuations governed by known probabilistic laws, and there is a very extensive literature on the subject. The mathematics has been presented in works by Gani (1957), Moran (1959), Prabhu (1964), Lloyd (1967), Kartvelishvili (1969), and Phartofod (1976, 1979, 1989). Excellent accounts for engineers have been given by McMahon and Mein (1978) and Klemes (1981), who gives a realistic evaluation of the limited contribution of the mathematical theory to engineering practice: not because of shortcomings in the theory itself, but because of uncertainties in the probabilistic laws governing the reservoir inflows. These probabilistic laws must of necessity be established from observations of the inflows over periods that are commonly short.

We begin by showing how Moran (1959) expressed reservoir behaviour, albeit of a simplified kind, in terms of (homogeneous) Markov chain theory. The input is assumed random, but the output depends upon the amount of water in the reservoir and on a "release rule" which prescribes how and when water is released from it. Time is assumed to be measured in years, and the release is assumed to occur just once a year at a specified date; this simplification may not be too unreasonable where the purpose of the reservoir is to store wet-season input for use in a pronounced dry season. Moran assumes:

1. That the total annual inflow X_t to the reservoir in year t is a random variable with continuous probability distribution defined on the interval $(0, \infty)$;
2. That the annual inflows X_t are independent of each other (the critical assumption);
3. That the X_t have the same probability distribution.

Let Z_t be the volume of water in the reservoir prior to the inflow X_t (say, at the end of the dry season, terminated by the onset of the wet); let K be the capacity of the reservoir; and let M be the "target" release, in the sense that a volume M is released in the dry season if, at its onset, the reservoir contains a volume greater than M. (If it does not, a volume smaller than M is released and the reservoir becomes empty.) During the wet season, the sum of the existing volume in storage Z_t is increased to $Z_t + X_t$, but if this exceeds the total reservoir volume K, the excess is lost as spillage before the target M is released. Hence the continuity condition for the reservoir content can be expressed as

$$Z_{t+1} = \begin{cases} 0 & (Z_t + X_t \leq M) \\ Z_t + X_t - M & (M < Z_t + X_t < K) \\ K - M & (K \leq Z_t + X_t). \end{cases}$$

Moran derives an integral equation (see next section) for the distribution of the random variable $Z_t + X_t$, given the distribution of the inflows X_t, but the equation can only be solved for special cases. As an alternative to the search for solutions of the integral equation, Moran suggests that

1. The reservoir volume K, and target release M, be "discretised" by expressing them both in terms of some unit of volume (say, $K = k$ times, and $M = m$ times, this selected unit);
2. That the inflows X_t be discretised in terms of the unit volume, so that the probability that $X_t = i$ is $P[X_t = i] = \pi_i$, say (with $\pi_0 + \pi_1 + \pi_2 + \ldots = 1$);
3. That the reservoir volume Z_t be discretised in terms of the unit volume, so that Z_t can take the discrete values $0, 1, \ldots k - m$. Writing $P_j(t)$ for the probability that the reservoir volume $Z_t = j$ at time t, so that $P_j(t) = P[Z_t = j]$, and $\boldsymbol{P}(t)$ for the row vector with $k - m + 1$ elements $P_0(t), P_1(t), \ldots P_{k-m}(t)]$, then

$$\boldsymbol{P}(t) = \boldsymbol{P}(t-1).\boldsymbol{\Pi} \tag{4.11}$$

where $\boldsymbol{\Pi}$ is the $(k - m + 1) \times (k - m + 1)$ matrix

	Col. 1	Col. 2	Col. 3	…	…	Col. $k-m$	Col. $k-m+1$
Row 1	$\pi_0 + \pi_1 + \ldots \pi_m$	π_{m+1}	π_{m+2}	…	…	π_{k-1}	$\pi_k + \pi_{k+1} + \ldots$
Row 2	$\pi_0 + \ldots \pi_{m-1}$	π_m	π_{m+1}	…	…	π_{k-2}	$\pi_{k-1} + \pi_k + \ldots$
Row 3	$\pi_0 + \ldots \pi_{m-2}$	π_{m-1}	π_m	…	…	π_{k-3}	$\pi_{k-2} + \pi_{k-1} + \ldots$
	…	…	…	…	…	…	
Row $m+1$	π_0	π_1	π_2	…	…	π_{k-m-1}	$\pi_{k-m} + \pi_{k-m+1} + \ldots$
Row $m+2$	0	π_0	π_1	…	…	π_{k-m-2}	$\pi_{k-m-1} + \pi_{k-m} + \ldots$
Row $m+3$	0	0	…				
	…	…	…				
Row $k-m+1$	0	0	0	…	…	π_{m-1}	$\pi_m + \pi_{m+1} + \ldots$

(4.12)

Some explanation is required concerning the form of the transition probabilities in the above matrix. Take, for example, the relation expressing $P_0(t)$, the probability that the reservoir is empty at time t, in terms of the probabilities $P_j(t-1)$ of being in the states $j = 0, 1, 2, \ldots k - m$ at time $t - 1$. Emptiness at time t can occur in a number of mutually exclusive ways: the reservoir may already have been empty (i.e. in state 0) at time $t - 1$, with the wet season input being less than or equal to m, this event having probability

$\pi_0 + \pi_1 + \ldots \pi_m$. An alternative is that the reservoir contained one unit of volume (was in state 1) at time $t - 1$, with wet-season input being less than or equal to $m - 1$ units of volume, this event having probability $\pi_0 + \pi_1 + \ldots \pi_{m-1}$; and so on, for the other alternatives that the reservoir was in states 2, 3,... at time $t - 1$. Since the alternatives are mutually exclusive, the probabilities of the alternatives are added, by the theory of Chapter Two, so that

$$\begin{aligned} P_0(t) &= P_0(t-1).[\pi_0 + \pi_1 + \ldots \pi_m] + P_1(t-1).[\pi_0 + \pi_1 + \ldots \pi_{m-1}] \\ &\quad P_2(t-1).[\pi_0 + \pi_1 + \ldots \pi_{m-2}] + \ldots \\ &= \boldsymbol{P}(t-1).[\pi_0 + \pi_1 + \ldots \pi_m \, \pi_0 + \pi_1 + \ldots \pi_{m-1} \ldots \pi_0 0 0 \ldots 0]^T \end{aligned}$$

where the transposed, square-bracketed vector is the first column of the matrix $\boldsymbol{\Pi}$. The zero elements occur because (for example) if the reservoir contains $m + 1$ units of volume at time $t - 1$, it cannot be empty at time t even if wet-season inflow is zero, since the target release of m units can still be met. Similarly, if the reservoir contains $m + 2, m + 3, \ldots k$ units of volume at time $t - 1$.

Equation (4.11) is the now-familiar equation giving the probabilities of being in state $0, 1, \ldots k - m$ at time t, in terms of the probabilities of the chain being in state $0, 1, \ldots k - m$ at time $t - 1$, and the transition probabilities between the states; the new element that has now been introduced is that the transition probabilities have now been expressed in terms of the (discretised) probability distribution of the reservoir inflows. The "steady-state" probabilities of the reservoir being in states $0, 1, 2, \ldots k - m$ as time t increases indefinitely are found by solving, for $\boldsymbol{P}$, the equation

$$\boldsymbol{P} = \boldsymbol{P}.\boldsymbol{\Pi} \tag{4.13}$$

subject to the condition that the $k - m + 1$ elements in the row vector $\boldsymbol{P}$ add to one; and remembering that the unit interval of time is one year, the probability of the reservoir containing volumes $0, 1, 2, \ldots k - m$ units of volume in year t is

$$\boldsymbol{P}(t) = \boldsymbol{P}(0).\boldsymbol{\Pi}^t \tag{4.14}$$

where $\boldsymbol{P}(0)$ is the row vector containing the probabilities of being in states $0, 1, 2, \ldots k - m$ at time $t = 0$ (if the reservoir is initially empty, $\mathbf{P}(0) = [1, 0, 0, \ldots 0]$, i.e. 1 with $k - m$ zeros).

The discrete solution given above, proposed by Moran (1959), was extended by a number of researchers, many of whom have preferred to modify Moran's original formulation. It will be seen in the above that although the reservoir contains k units of volume, its contents never exceed $k - m$ units; this is because the volume Z_t at the end of the dry season is increased by wet season inflow X_t, before the dry season target M is released (or before the entire contents are released, in the case when $Z_t + X_t$ is less than M). This is appropriate where, as in parts of Australia for which Moran developed his model, a reservoir is fed by a single brief Nile-like flood. An alternative formulation (see e.g. McMahon and Mein, 1978) assumes that inflows and outflows occur simultaneously, such that the inflow X_t, and the release with target value M, are spread uniformly over the working interval $(t, t+1)$. (If, however, the inflow rate is so small that the reservoir becomes empty during the interval, the release rate is set equal to the inflow rate, so that the reservoir remains empty until the end of the interval; whilst if the inflow rate is so large that the reservoir becomes full before the end of the interval, it remains full, and spilling, until the end of that interval.) With this convention of "simultaneous" inflows and releases, the continuity conditions are slightly modified to become

$$Z_{t+1} = \begin{cases} 0 & (Z_t + X_t \leq M) \\ Z_t + X_t - M & (M < Z_t + X_t < K + M) \\ K & (K + M \leq Z_t + X_t) \end{cases} \qquad (4.15)$$

and values of Z_t now satisfy $0 \leq Z_t \leq K$, so that reservoir volume can assume values from zero to its maximum value K.

With this convention of simultaneous inflow and release, Kaczmarek (1963), Lloyd (1963) and Phatarfod and Mardia (1973) relaxed the assumption that the reservoir inflows be statistically independent, by permitting them to be serially correlated. Other studies (Prabhu, 1958; Phatarfod, 1979; Lloyd and Warren, 1992) explored special methods for solving the linear equations. Lloyd (1993) points out that much of this earlier work can be criticised because it paid insufficient attention to the errors involved in "discretising" the inflows and storage levels and, depending heavily on Lloyd's penetrating paper, we return to this problem in section 4.6.

4.5.1 Numerical example

Let us suppose that the annual inflows X_t to a reservoir take the values $1, 2, 3, \ldots 10$ units of volume with probabilities

$$\begin{array}{ll} P[X_t = 1] = \pi_1 = 0.264241 & P[X_t = 2] = \pi_2 = 0.329753 \\ P[X_t = 3] = \pi_3 = 0.206858 & P[X_t = 4] = \pi_4 = 0.107570 \\ P[X_t = 5] = \pi_5 = 0.051150 & P[X_t = 6] = \pi_6 = 0.023076 \\ P[X_t = 7] = \pi_7 = 0.010056 & P[X_t = 8] = \pi_8 = 0.004276 \\ P[X_t = 9] = \pi_9 = 0.001785 & P[X_t = 10] = \pi_{10} = 0.000735\ldots. \end{array}$$

Suppose that the reservoir capacity is 5 units of volume, and that the target release is 3 units. We take 6 reservoir states, numbered 0 to 5, corresponding to empty (state 0); reservoir volume between 0 and 1 unit (state 1); reservoir volume between 1 and 2 units (state 2); between 2 and 3 units (state 3); between 3 and 4 units (state 4); and reservoir full (state 5). The matrix $\mathbf{\Pi}$ of transition probabilities is then

$$\begin{bmatrix} 0.800852 & 0.107570 & 0.051150 & 0.023076 & 0.010056 & 0.007296 \\ 0.593994 & 0.206858 & 0.107570 & 0.051150 & 0.023076 & 0.017352 \\ 0.264241 & 0.329753 & 0.206858 & 0.107570 & 0.051150 & 0.040429 \\ 0 & 0.264241 & 0.329753 & 0.206858 & 0.107570 & 0.091579 \\ 0 & 0 & 0.264241 & 0.329753 & 0.206858 & 0.199149 \\ 0 & 0 & 0 & 0.264857 & 0.329753 & 0.406006 \end{bmatrix}$$

If we suppose that the reservoir is initially empty ($\mathbf{P}(0) = [1, 0, 0, 0, 0, 0]$) then the multiplication in (4.14) gives

$$\begin{aligned} \mathbf{P}(1) &= [0.801 \quad 0.107 \quad 0.051 \quad 0.023 \quad 0.010 \quad 0.070] \\ \mathbf{P}(2) &= [0.719 \quad 0.131 \quad 0.073 \quad 0.040 \quad 0.020 \quad 0.017] \\ \mathbf{P}(3) &= [0.673 \quad 0.139 \quad 0.084 \quad 0.050 \quad 0.028 \quad 0.024] \\ \mathbf{P}(4) &= [0.644 \quad 0.142 \quad 0.091 \quad 0.058 \quad 0.034 \quad 0.031] \\ \mathbf{P}(5) &= [0.642 \quad 0.144 \quad 0.095 \quad 0.063 \quad 0.038 \quad 0.035] \\ \mathbf{P}(10) &= [0.586 \quad 0.147 \quad 0.103 \quad 0.074 \quad 0.046 \quad 0.044] \\ & \qquad \qquad \ldots \qquad \qquad \ldots \\ \mathbf{P}(\infty) &= [0.578 \quad 0.147 \quad 0.105 \quad 0.076 \quad 0.048 \quad 0.046] \end{aligned}$$

where $\mathbf{P}(\infty)$ gives the steady-state probabilities found by solving (4.13). If, instead of multiplying matrices in (4.14) to obtain $\mathbf{P}(t)$, we had used the decomposition $\mathbf{\Pi} = \mathbf{Q\Lambda Q}^{-1}$ where $\mathbf{\Lambda}$ is the matrix having the eigenvalues of $\mathbf{\Pi}$ on its principal diagonal and $\mathbf{Q}$ is the matrix having eigenvectors as its columns, the probabilities in $\mathbf{P}(10)$ agree to the fourth decimal place; the matrix $\mathbf{\Lambda}$ is

$$\mathbf{\Lambda} = \text{diag}[1 \quad 0.706747 \quad 0.277505 \quad 0.048117 \quad 0.001920 \quad 0]$$

and the matrix $\mathbf{Q}$ is, to five decimal places,

$$\mathbf{Q} = \begin{bmatrix} 0.40825 & -0.13397 & 0.04325 & 0.00315 & -0.00002 & 0.00000 \\ 0.40825 & -0.06856 & -0.07130 & -0.04240 & 0.00292 & 0.00000 \\ 0.40825 & 0.05754 & -0.24706 & -0.06071 & -0.01738 & 0.00000 \\ 0.40825 & 0.25970 & -0.35616 & 0.17632 & -0.14954 & -0.18442 \\ 0.40825 & 0.51959 & -0.06339 & 0.65353 & 0.81158 & 0.81822 \\ 0.40825 & 0.79789 & 0.89506 & -0.73234 & -0.56450 & -0.54452 \end{bmatrix}.$$

An alternative to the solution of (4.13) and (4.14) is Monte Carlo simulation (a method which almost certainly would need to be used to calculate steady-state probabilities in an interconnected series of reservoirs).

4.5.2 The Jerimum reservoir of north-east Brazil

This example has been kindly provided by Eduardo Sávio Martins. The

Jerimum reservoir was planned to impound the River Caxitoré in the Brazilian State of Ceará, in the semi-arid north-east of the country. The catchment area upstream of the proposed reservoir site is 386 km^2 with an altitude range from 612 m to 130 m; the reservoir was designed to reduce water loss to the sea and to maintain an irrigated area of about 300 km^2. Mean annual runoff for the period 1959–90 was 124.7×10^6 m^3, but there was no runoff at all in either of the years 1982–83, whilst in 1974 total runoff was 902.7×10^6 m^3. The period September to December is dry in every year; in the period of record, no runoff was recorded in October and December, with a small volume of runoff in November in two years of the 32.

The site of the Jerimum reservoir corresponded closely with the kind of environment for which Moran had developed storage theory. A mixed rock and earth dam was planned with a capacity of 20.5×10^6 m^3. An annual time-step was used, and the reservoir volume was divided into 20 discrete intervals. Serial correlation between annual inflows was taken as zero, all inflows were assumed to occur in the wet season, and all outflows in the dry season. Output was taken to consist of a three-part sequence: initially half the evaporative loss was deducted, then the total planned release, and finally the second half of the evaporative loss. This division of the evaporative loss was intended to secure a "mean value" for evaporation loss, in the sense that its first part is deducted when the reservoir level is high (and before the planned release), whilst the second part is deducted with the reservoir at a lower level (after release). A final assumption was that annual inflows followed a two-parameter gamma distribution, with a discrete particle of probability at the origin to allow for zero inflows.

Evaporation loss is a function of evaporative demand (for example, Penman's open-water evaporation) and extent of water surface, so the volume evaporated depends on local climatic conditions and reservoir geometry. In designing the Jerimum reservoir, the following relationships were used to relate reservoir storage Z to area of water surface A and height of water surface H:

$$Z(H) = \alpha H^3$$
$$A(H) = \mathrm{d}Z/\mathrm{d}H = 3\alpha H^2$$

where α is a parameter of reservoir shape, determined by least squares from data relating $Z(H)$ to H^3. Denoting by E_o the open-water evaporation during any period, the evaporated volume V when the reservoir is at level H is then $A(H).E_o$, or

$$V = 3\alpha^{1/3}.E_o.[Z(H)]^{2/3}.$$

A transition matrix **E** was then defined, with element e_{ij} calculated from the values V, as follows. With a the increment of reservoir storage, the reservoir content in state i is $i.a$, and the reduction in volume due to an evaporative loss V is approximately $N_L = i.a. - [V/a]$ where [.] here signifies "integral part of". Then for the ith row of **E**, if $i > N_L, e_{ij}$ was put equal to one for $j = i - N_L$, and zero otherwise, whilst if $i \le N_L, e_{ij}$ was set to one for $i = 0$, and zero otherwise.

With the matrix **E** so defined, another transition matrix **W** was defined for the inflows. The matrix **W**, with element w_{ij}, was upper triangular, since inflows must be non-negative so that with the reservoir in state $i, w_{ij} = 0$ for $j < i$. A similar matrix **R** was defined for the releases. The vector of probabilities $\boldsymbol{P}(t)$ that the reservoir was in state i at time t was then defined by the relation

$$\boldsymbol{P}(t+1) = \mathbf{E.R.E.W.P}(t).$$

The approach supposes that evaporation loss is constant and independent of t, but it would be straightforward to extend it to the case where annual evaporation has a probability distribution; the rows of the matrix **E** would then contain elements other than zero or one.

4.6 THE INTEGRAL EQUATION FOR RESERVOIR STORAGE

We now consider the case where Z_t and X_t are allowed to vary continuously over the intervals $(0, K)$ and $(0, \infty)$ respectively. It can be shown that the probability distribution of Z_t then has a concentration of probability at 0 and K, and is continuous between these values: in other words, there is a discrete particle or mass of probability in each of the states 0 and K, with the remaining mass spread continuously in between, the total of the two particles and the continuous part summing to one. The objective is to derive the probability distribution of Z_t. The masses of probability at 0 and K cause an inconvenience which can be avoided by working not with the probability distribution of Z_t directly, but with the probability distribution of the sum $U_t = Z_t + X_t$, from which the distribution of Z_t can subsequently be derived. Write $h_t(u)$ for the probability distribution of U_t (so that $h_t(u)\mathrm{d}u = P[u < U_t < u + \mathrm{d}u]$ with $u \ge 0$) and $f(x)$ for the probability distribution of $X_t, f(x)\mathrm{d}x = P[x < X_t < x + \mathrm{d}x]$ also for $x \ge 0$. The equations relating U_t to inflow X_t then become

$$U_{t+1} = \begin{matrix} X_{t+1} & & U_t \le M \\ U_t + X_{t+1} - M & M < & U_t < K + M \\ X_{t+1} + K & & K + M \le U_t \end{matrix} \tag{4.16}$$

Then, by the theorem of total probability of Chapter Two

$$h_{t+1}(u) = \int_{v=0}^{\infty} h_{t+1}(u \mid U_t = v) h_t(v) \mathrm{d}v$$

$$= \int_{v=0}^{M} h_{t+1}(u \mid U_t = v)h_t(v)\mathrm{d}v + \int_{v=M}^{K+M} h_{t+1}(u \mid U_t = v)h_t(v)\mathrm{d}v$$

$$+ \int_{v=K+M}^{\infty} h_{t+1}(u \mid U_t = v)\mathrm{h}_t(v)\mathrm{d}v$$

on dividing the range of integration into the sub-ranges from 0 to M, M to $K + M$, and $K + M$ to ∞as used in (4.16). But because of (4.16), the conditional probability $h_{t+1}(u \mid U_t = v)$ in the first integral equals $f(u)$, whilst in the second and third integrals it equals $f(u + M - v)$ and $f(u - K)$ respectively, so that

$$h_{t+1}(u) = f(u) \int_{v=0}^{M} h_t(v)\mathrm{d}v + \int_{v=M}^{K+M} h_t(v)f(u + M - v)\mathrm{d}v$$

$$+ f(u - K) \int_{v=K+M}^{\infty} h_t(v)\mathrm{d}v \qquad (4.17)$$

which is the integral-difference equation for the reservoir, using the simultaneous inflow and release convention. If, instead of working with the probability distributions $h_t(u)$ and $f(x)$, we work with cumulative distribution functions

$$H_t(u) = \int_{w=0}^{u} h_t(w)\mathrm{d}w \quad \text{and} \quad F(x) = \int_{w=0}^{x} f(w)\mathrm{d}w$$

then (4.17) may be written in an alternative form

$$H_t(u) = F(u - K) - \int_{w=M}^{K+M} H_{t-1}(w)f(\mathrm{u} + M - w)\mathrm{d}w. \qquad (4.18)$$

If, as $t \to \infty$, the cumulative distribution function $H_t(u)$ tends to a limiting form $H(u)$, say, then (4.17) becomes

$$H(u) = F(u - K) - \int_{w=M}^{K+M} H(w)f(\mathrm{u} + M - w)\mathrm{d}w. \qquad (4.19)$$

An equivalent equation in terms of probability density functions $h(u)$ and $f(x)$ is obtained by dropping the suffices $t, t + 1$ from (4.17).

We recall at this point that the probability distribution that is really of interest is not that of $U_t = Z_t + X_t$, but of Z_t, the reservoir contents at time t.

We can derive this distribution once the distribution of U_t is calculated from (4.17), because

$$\begin{aligned} P[Z_{t+1} = 0] &= P[Z_t + X_t \leq M] \\ &= P[U_t \leq M] \\ &= H_t(M) \end{aligned} \tag{4.20}$$

and also

$$P[Z_{t+1} = K] = P[U_t \geq K + M] = 1 - H_t(K + M). \tag{4.21}$$

Thus the two particles of probability at 0 and K can be established once $h_t(u)$ – and hence $H_t(u)$ – is known. Within the open interval $(0, K)$, the probability distribution of Z_t is continuous; but by (4.14) we have

$$Z_{t+1} = U_t - M \quad \text{for } M < U_t < K + M$$

and it follows that the distribution of Z_{t+1} between 0 and K is

$$g_{t+1}(z) = h_t(z + M) \quad \text{for } 0 < z < K. \tag{4.22}$$

Equations (4.20), (4.21) therefore give the masses of probability at 0 and K, whilst (4.22) gives the continuous distribution of probability within the interval $(0, K)$.

Although most of the problem is now solved, there are still some tricky points that arise when probability distributions like that of Z_{t+1} are discretised. Equations (4.11) and (4.13) were given without considering how the discretisation should be effected; one procedure has been to divide the reservoir volume K into the $k + 1$ states

$$\begin{aligned} &0 \leq Z_{t+1} < \tfrac{1}{2}; \tfrac{1}{2} \leq Z_{t+1} < 3/2; \ldots K - 3/2 \\ &\leq Z_{t+1} < K - \tfrac{1}{2}; K - 1/2 \leq Z_{t+1} \leq K, \end{aligned}$$

known as Moran's discretisation. However since the distribution of Z_{t+1} has two masses of probability at $Z_{t+1} = 0$ and $Z_{t+1} = K$, it might be expected that a better way of discretising the closed interval $[0, K]$ would be in terms of the $k + 2$ intervals (corresponding to $k + 2$ states) defined by

$$Z_{t+1} = 0; 0 < Z_{t+1} \leq 1; 1 < Z_{t+1} \leq 2; \ldots K - 1 < Z_{t+1} < K; Z_{t+1} = K.$$

The reader is referred to the paper by Lloyd (1993) for an account of how the alternative subdivisions of $[0, K]$ influence numerical solutions for the probability distribution $g_{t+1}(z)$ of (4.22).

We return to (4.17) and (4.18) in the next chapter, where we consider the modelling of soil moisture storage at a river basin scale.

4.7 CONCLUSION

Some of the mathematics of this section is daunting when first encountered. However, the reader who is more concerned with solutions to practical problems of reservoir design and operation should not be disturbed by the complexities of integral-difference equations like (4.16); Lloyd (1993) says

> the so-called stochastic reservoir must be distinguished from its real counterpart. The latter is a complex multi-purpose engineering construction, possibly part of an interconnected network, in which the water level is subject not only to irregular natural inflows – possibly with controlled supplements – and managed releases, but also to losses through seepage, evaporation, wind-blown spray, and so forth. . . . The stochastic reservoir, on the other hand, is a mathematical model of a highly simplified version of this. It is a representation of an isolated container receiving inflows that are subject to random fluctuations governed by known probabilistic laws.

Where complicated interconnected reservoir systems are to be designed, or their operating rules determined, opportunities for utilising the mathematical theory of storage will be the exception rather than the rule, and it is more likely that simulation of inflow sequences, following the application of time series analysis to determine an appropriate form for the cross- and serial-correlation structures, will prove the more practical approach. Nevertheless a mathematical theory, even if greatly simplified, is particularly useful in that it provides insight into a problem, a check that a simulation study gives conclusions of the right order, and a starting-point for more complex theories with fewer and more realistic assumptions.

Chapter Five

Stochastic reservoir theory applied to rainfall–runoff modelling

5.1 SOME BASIC IDEAS IN RAINFALL–RUNOFF MODELLING

This chapter seeks to utilise aspects of storage theory shown in preceding chapters, to the modelling of rainfall–runoff behaviour.

A rainfall–runoff model is a device, either mathematical, computational or analogue, for expressing how river runoff responds to precipitation, which may be either a spatial average over a drainage basin upstream of a site at which river flow is calculated, or the "point" precipitation measured at several sites within the basin. Many models also use meteorological variables from which evaporation (essentially regarded as a negative input) can be estimated. One very simple model regards a drainage basin as a "linear reservoir" for which the precipitation input (or some variable derived from it, such as "effective" precipitation, or "precipitation minus potential evaporation") is $I(t)$, and the runoff leaving the basin is $Q(t)$, with $Q(t)$ proportional to $S(t)$, the volume currently stored within the reservoir. Thus the equations describing this very simple system are

$$\mathrm{d}\,S(t)/\mathrm{d}t = I(t) - Q(t) \tag{5.1}$$

which is the continuity equation, and $Q(t) = K.S(t)$, expressing the linearity of the system. (A non-linear reservoir may, for example, be given by $Q(t) = K.S(t)^n$, where n is some power other than one.) Hence

$$\mathrm{d}S(t)/\mathrm{d}t + KS(t) = I(t) \tag{5.2}$$

which can be put in discrete terms as

$$S_{t+1} = (1 - K.\Delta t)S_t + I_t.\Delta t. \tag{5.3}$$

The same equations can be expressed in terms of runoff $Q(t)$, instead of $S(t)$, using $Q(t) = K.S(t)$. Thus (5.2) can be written as an equation in $Q(t)$:

$$K^{-1}\mathrm{d}Q(t)/\mathrm{d}t + Q(t) = I(t).$$

Equations (5.2) and (5.3) can be regarded as deterministic; however, if S_t and $I_t.\Delta t$ are regarded as random variables, (5.3) has the same form as the "lag-one autoregression model" (AR(1)) familiar to time-series analysts, since it can be written

$$S_{t+1} - \mu_s = (1 - K.\Delta t)(S_t - \mu_s) + (I_t - K.\mu_s).\Delta t \tag{5.4}$$

expressing the deviation of S_{t+1} from its mean value μs in terms of the corresponding deviation at time t, and a random disturbance $(I_t - K.\mu s).\Delta t$. Note that if (5.4) is regarded as a stationary autoregressive model, $K, \Delta t$ must satisfy $0 < K.\Delta t < 2$; and unless the time interval is large – say, a year – the random disturbances $\varepsilon_t \equiv (I_t - K.\mu s).\Delta t$ are unlikely to satisfy the usual conditions $\mathrm{E}[\varepsilon_t] = 0, \mathrm{E}[\varepsilon_t^2] = \sigma^2$, a constant, and $\mathrm{E}[\varepsilon_t\varepsilon_{t+m}] = 0$ for $m \neq 0$ required for a valid AR(1) time series model.

Subject to certain conditions, the linear reservoir model may also be expressed in terms of probability distributions. If (1) the input $I_t.\Delta t$ is spatially averaged over the drainage basin, and (2) this input is regarded as a serially uncorrelated random variable with a probability distribution "discretised" into intervals of depth denoted by a, (3) the input is statistically independent of the storage state S_t, then the linear reservoir model can be regarded as a Markov chain with an infinite number of states. For we have

$$P[S_{t+1} = ka] = \sum_{l=0}^{\infty} P[S_t = la].P[I_t.\Delta t = (k - l)a] \tag{5.5}$$

where, for practical use, the infinite upper limit would be replaced by some appropriate large number M.

Hence, the linear reservoir model can be regarded as a deterministic description of river basin behaviour; or, in discretised form and subject to certain conditions, as a time-series model of river basin behaviour; or again subject to certain conditions, as a Markov chain model of river basin behaviour. However, the linear reservoir would be a greatly oversimplified description, and the conditions necessary for the AR(1) or Markov-chain representations will not be satisfied for small Δt. Nevertheless the linear reservoir is often used as one component in more complex rainfall–runoff models, in which storages are linked in series and in parallel. These more complex models usually require knowledge of parameters (like the constant K of proportionality in the linear reservoir model) which are commonly estimated by minimising a criterion such as the sum of squared differences between observed and predicted series of runoff values Q_t. Various practical difficulties, such as failure to converge, and the presence of several minima, are commonly encountered where such criteria are minimised.

5.2 THE USE OF STORAGE THEORY IN RAINFALL–RUNOFF MODELLING

Field observation shows that drainage basins store water in various ways: in surface depressions, on vegetation and in soil. Furthermore, the "threshold" at which storage becomes full and begins to overflow varies, often greatly, from one part of the basin to another. Many models of basin response represent such components of storage by a single reservoir having a single threshold at which spillage begins, but the presence of such single thresholds causes some of the complications encountered when parameters are estimated by least squares, mentioned in section 5.1. A way of avoiding some of these problems was suggested by Moore and Clarke (1981), who found that the sum of squared errors surfaces became smoother, so that their mimima could be reached more easily, when the single reservoir was replaced by a population of reservoirs, each of which is characterised by a different threshold at which spillage begins; the depth of reservoir storage in the population of reservoirs then has a statistical distribution. We denote the capacity of storage in any particular reservoir by S, so that at any instant the depth of water stored in it lies between zero and S.

The model of Moore and Clarke (1981), subsequently discussed also by Hosking and Clarke (1989, 1991), is as follows. At any point P in the drainage basin, at which there is a hypothetical storage of capacity S, assume that runoff occurs only when the depth of water in storage at P exceeds S; that is, when the soil moisture deficit, denoted by X, has been restored to zero. Once this state is reached, any further rainfall becomes runoff. After rainfall stops, water evaporates at a potential rate determined by meteorological conditions, thereby re-creating soil moisture deficit X, and evaporation continues at the potential rate until the capacity S at the point P has been completely exhausted; evaporation at the point P then stops (although it will continuc from other points P*, with capacities S^*, say, which are greater than S, and which have not yet become exhausted).

Suppose now that the distribution of the storage capacity S is

$$P[s < S < s + \mathrm{d}s] = p(s; \boldsymbol{\theta})\mathrm{d}s \qquad 0 \leq S < \infty,$$

where $\boldsymbol{\theta}$ is a vector of parameters. If the total basin area is A, then a proportion $A.p(s;\theta)\mathrm{d}s$ has capacity for storage between $(s, s + \mathrm{d}s)$. Clearly some explicit form must be specified for $p(s;\theta)$, and the parameters $\boldsymbol{\theta}$ will then determine the mean and variance (and other moments) of storage capacity across the basin as a whole. If $p(s;\boldsymbol{\theta})$ consists merely of a unit mass of probability at some value

$S = S_0$, say, then the population of storages reduces to the single storage that is a typical component of "lumped" rainfall–runoff models.

Moore (1985) gives a historical review of the concept of probability distributed storage, showing that although not explicitly stated, the concept underlay much published research extending back to the 1950s. Moore traces its usage, citing Russian research by Bagrov (1950), Popov (1956, 1962, 1973, 1980) and work by the East China College of Hydraulic Engineering (Zhao and Zhuang, 1963; Zhao et al. 1980) and, in the former GDR, Becker (1974).

5.2.1 An example: exponential distribution of storage capacity

We take as an example the case where storage capacity S has an exponential distribution with parameter $p(s; \theta)$:

$$p(s; \theta) = \theta \exp(-\theta s). \tag{5.6}$$

Suppose that at time $t = 0$, all storage capacity has been filled, and that evaporation then begins at a constant rate E. At time $t > 0$, all elements of storage for which $S < E.t$ will be empty, their contents having evaporated; however, elements for which $S > E.t$ will continue to evaporate at the potential rate E. The mean rate of *actual* evaporation across the basin will therefore be

$$E_a = E. \int_{s=Et}^{\infty} \theta \exp(-\theta s) \mathrm{d}s \tag{5.7}$$

giving $E_a/E = \exp(-Et)$, so that mean actual evaporation declines exponentially from its initial potential value E as the length of dry period extends. A similar calculation gives the mean deficit in storage at time t: it is

$$SMD = \int_{s=0}^{Et} s\theta \exp(-\theta s) \mathrm{d}s + \int_{s=Et}^{\infty} Et.\theta \exp(-\theta s) \mathrm{d}s$$

giving

$$SMD = \theta^{-1}\{1 - \exp(-Et)\} + Et \exp(-Et).(1 - \theta^{-1}). \tag{5.8}$$

As $t \to \infty$, SMD therefore tends to θ^{-1}, the mean value of the exponential distribution of S across the drainage basin, which equals the mean depth available for the storage of soil moisture. The two expressions (5.8) and $E_a/E = \exp(-Et)$, obtained from (5.7), can be used to give an explicit relation between mean actual evaporation rate E_a and mean soil moisture deficit SMD; by eliminating t, we obtain:

$$SMD - \theta^{-1} = -(E_a/E).\{(1 - \theta^{-1})\log_e(E_a/E) + \theta^{-1}\}. \tag{5.9}$$

If $\theta = 1$, for example, this relation takes a very simple form

$$E_a/E = 1 - SMD$$

confirming that actual evaporation is at the potential rate ($E_a = E$) when $SMD = 0$, and that $SMD = 1$ when $E_a = 0$. For other values of θ, (5.9) takes a more complicated form, but can be approximated by

$$SMD - \theta^{-1} = -(E_a/E).\{(1 - \theta^{-1})((E_a/E - 1) - 1/2(E_a/E - 1)^2 + 1/3(E_a/E - 1)^3 - \ldots) + \theta^{-1}\}.$$

For a general distribution of capacity $p(s;\theta)$, the mean actual evaporation E_a is given by

$$E_a = E \int_{s=Et}^{\infty} p(s;\theta)\mathrm{d}s$$

and the mean storage deficit is

$$SMD = \int_{s=0}^{Et} sp(s;\theta)\mathrm{d}s + Et \int_{s=Et}^{\infty} p(s;\theta)\mathrm{d}s.$$

Thus it is seen that the advantage of using a distribution $p(s;\theta)$ to describe soil moisture storage within a drainage basin is that it removes the awkward "elbows" or thresholds in piecewise descriptions of the relationship between actual evaporation loss and mean areal soil moisture deficit. The price to be paid is that the distribution of storages $p(s;\theta)$ must be defined, and its parameters estimated; the relation (5.9) relating actual evaporation to soil moisture deficit will have a different form if the distribution $p(s;\theta)$ is gamma, inverse-normal or of some other form, so its choice is critical.

Moore and Clarke (1981) show diagrammatically how the disposition of empty, partially filled and completely filled storages varies across a drainage basin for general choice of $p(s;\theta)$, as rainfall events develop and are followed by dry periods during which soil moisture is depleted. Readers of Chapter Six will recognise that the model described by Moore and Clarke (1981) is essentially an alternating renewal process for each element of storage with depth S in the range $(s, s + \mathrm{d}s)$. Associated with the alternating renewal process are two "mark" state variables: namely, the runoff (spillage) generated by each rainfall event, and the depth of water remaining in storage.

5.3 RUNOFF ROUTING

Up to this point, only the representation of basin storage has been considered. A drainage basin also transmits whatever runoff is generated within it to the basin outfall. Suppose now that the basin is at the end of a very long dry period, with all water evaporated from storage so that the mean deficit is at its maximum, with actual evaporation at its minimum. If, at time $t = 0$, it now begins to rain at a constant rate P, say, than at time t up to which a depth of rain $P.t$ has fallen, a depth of runoff will have been generated that is equal to

$$\int_{s=0}^{Pt} (Pt - s)p(s;\theta)\mathrm{d}s.$$

In the case where $p(.)$ is the exponential distribution used in section 5.2.1, the generated runoff can be shown to be

$$Pt[1 - \exp(-Pt)] + \theta^{-1}[Pt\exp(-Pt) - (1 - \exp(-Pt))]$$

which tends to infinity as t tends to infinity; thus after continuous steady rain has started, the depth of generated runoff grows continually larger, and is generated over an increasingly wider area, as time passes.

Suppose now that the runoff generated at a point P within the basin translates to its outfall in time T. Moore and Clarke (1981) regarded the translation time T as coming from a probability density $p(t, \phi)$, so that

$$P[t < T < t + \mathrm{d}t] = p(t, \phi)\mathrm{d}t.$$

The density functions of storage, $p(s;\theta)$, and of translation, $p(t, \phi)$, are of course not necessarily the same. Furthermore, in the description given here the functions of basin storage and runoff routeing have been regarded as separable. It would be possible to treat them together, using a bivariate storage-translation function $p(s, t, \theta, \phi)$. Here

$$p(s, t, \theta, \phi)\mathrm{d}s\mathrm{d}t = P[(s < S < s + \mathrm{d}s) \cap (t < T < t + \mathrm{d}t)]$$

represents the proportion of the drainage basin with storage capacity lying between $(s, \mathrm{d}s)$ for which generated runoff is translated to appear at the basin outfall after a time lying between $(t, \mathrm{d}t)$.

In the separable case, the function $p(t, \phi)$ will be recognised as the instantaneous unit hydrograph. In the examples given above, it was convenient to assume that potential evaporation E and rainfall intensity P were constant, but of course both E and P will be time-variant. Moreover, wet and dry periods alternate, and in the next section we begin to incorporate such difficult realities.

5.4 DISTRIBUTIONS OF STORAGE CAPACITIES AND TRANSLATION TIMES COMBINED WITH RAINFALL CHARACTERISTICS

Earlier chapters have discussed the modelling of rainfall sequences using Markov chains to represent the alternation of wet and dry spells. We now use a two-state, first-order Markov chain, with α the probability that the $(i + 1)$th time interval Δt is wet given that the ith interval is dry, with β the probability that the $(i + 1)$th time interval is dry given that the ith time interval was wet. Denoting the two states (dry, wet) by (0, 1), the matrix of transition probabilities is

		$(i+1)$th interval: 0	1
ith interval:	0	$1-\alpha$	α
	1	β	$1-\beta$

For mathematical convenience it is assumed here that α, β are time-invariant.

Regarding the depths of rainfall in successive intervals Δt of a wet spell, these are assumed to be serially independent. If U is the random variable denoting depth of rainfall, the distribution function $P[U \leq u]$ is denoted by $W(u)$, and the probability density function $\mathrm{d}W(u)/\mathrm{d}u$ is denoted by $w(u)$. Similarly, the potential evapotranspiration in the successive intervals Δt of a dry spell are taken to be serially independent with common distribution function; with V denoting potential evapotranspiration, the distribution function $\mathrm{P}[V \leq v]$ is $D(v)$, and the probability density function $\mathrm{d}D(v)/\mathrm{d}v$ by $\mathrm{d}(v)$.

Now let J_i be a random variable taking values 0 or 1, according to whether the ith time interval Δt is dry or wet. For elements of storage with depth S lying in the interval $(s, s + \mathrm{d}s)$, let X_i denote the *SMD* at the end of the ith interval Δt. Thus X_i is the depth available in such elements of storage to accommodate all or part of any net rainfall occurring during the $(i + 1)$th interval Δt. The evolution of X_i is then governed by the recursion

$$\begin{aligned} X_{i+1} &= \max\{X_i - D_{i+1}, 0\} \qquad J_{i+0} = 0 \\ X_{i+1} &= \min\{X_i + W_{i+1}, s\} \qquad J_{i+1} = 1. \end{aligned}$$

Let

$$\begin{aligned} F_i^D(x; s) &= P[X_i \leq x \mid J_i = 0; s < S \leq s + \mathrm{d}s] \\ F_i^W(x; s) &= P[X_i \leq x \mid J_i = 1; s < S \leq s + \mathrm{d}s]. \end{aligned}$$

To simplify the notation, we omit for the time being the dependence on s so that these two definitions become

$$\begin{aligned} F_i^D(x) &= P[X_i \leq x \mid J_i = 0] \\ F_i^W(x) &= P[X_i \leq x \mid J_i = 1]. \end{aligned} \tag{5.10}$$

Then the distribution function of X_i, $P(X_i \leq x] = F_i(x)$ say, can be expressed in terms of $F_i^D(x)$ and $F_i^W(x)$ by using the theorem of total probability:

$$F_i(x) = F_i^D(x).P[J_i = 0] + F_i^W(x).P[J_i = 1]. \tag{5.11}$$

Now consider $F_{i+1}^{D}(x)$. We can obtain a recursive relation expressing it in terms of $F_i^D(x)$ and $F_i^W(x)$, as follows. Suppose that the ith period was dry ($J_i = 0$), and that the probability density of X_i at the end of the period is $d[F_i^D(x)]$; then if the $(i + 1)$th period is also dry, the probability that X_{i+1} is less than or equal to x must be the sum of terms like $D(x - y).d[F_i^D(y)]$, summed over all possible y, because soil moisture deficit X_i at the end of the ith period, and potential evaporation in the $(i + 1)$th period are independent so that the probabilities multiply. But this sum of terms $D(x - y).\mathrm{d}[F_i^D(y)]$ must be multiplied by the probability of the independent event of a transition from $J_i = 0$ to $J_{i+1} = 0$, namely $(1 - \alpha)$. By a similar argument for the case in which the ith period was wet, we obtain the equation

$$F_{i+1}^{D}(x) = P[X_{i+1} \leq x \,|\, J_{i+1} = 0]$$

$$= (1 - \alpha)\int D(x - y)\mathrm{d}F_i^D(y) + \alpha \int D(x - y)\mathrm{d}F_i^W(y). \qquad (5.12)$$

Similarly for $F_{i+1}^{W}(x)$: we have

$$F_{i+1}^{W}(x) = P[X_{i+1} \leq x \,|\, J_i = 1]$$

$$= \beta \int \{1 - W(y - x)\}\mathrm{d}F_i^D(y) + (1 - \beta)\int \{1 - W(y - x)\}\mathrm{d}F_i^W(y). \qquad (5.13)$$

Integrating the two equations (5.12) and (5.13) by parts, and using appropriate boundary conditions:

$$F_i^D(x) = F_i^W(x) = 0 \quad \text{for } x < 0 \quad \text{and} \quad F_i^D(x) = F_i^W(x) = 1 \quad \text{for } x > s,$$

Hosking and Clarke (1990) obtain the following equations:

$$F_{i+1}^{D}(x) = \int_0^x \{1 - \alpha)F_i^D(y) + \alpha F_i^W(y)\}d(x - y)\mathrm{d}y \qquad (5.14)$$

$$1 - F_{i+1}^{W}(x) = \int_{y=x}^{s} [\beta\{1 - F_i^D(y)\} + (1 - \beta)\{1 - F_i^W(y)\}w(y - x)\mathrm{d}y \qquad (5.15)$$

where $d(x - y), w(x - y)$ are the probability density functions of potential evaporation and rainfall in the time intervals Δt.

Thus (5.14) and (5.15) enable us to determine the conditional distribution functions $F_i^D(y)$ and $F_i^W(y)$, at any time, given (1) the distribution function at time $t = 0$, (2) the distribution functions $D(x)$ and $W(x)$ of potential evaporation and rainfall in dry and wet intervals respectively (both assumed serially independent); and (3) the transition probabilities α and β from dry to wet and wet to dry intervals Δt respectively. The unconditional distribution function of soil moisture deficit *SMD* is then given by

$$F_i(x) = F_i^D(x).P[J_i = 0] + F_i^W(x).P[J_i = 1]. \tag{5.16}$$

Recall, however, that we had dropped the variable s from $F_i^D(x)$ and $F_i^W(x)$, for notational simplicity, so that (5.14), (5.15) and (5.16) all refer to an element of basin storage of arbitrary depth s. To obtain a the basin-wide distribution function of *SMD*, it is necessary to multiply (5.16) by $p(s; \theta)$ and integrate with respect to s.

We can obtain "equilibrium" expressions for *SMD* by using the results, obtained previously, that for large i when initial conditions are no longer effective,

$$P[J_i = 0] = \beta/(\alpha + \beta), \qquad P[J_i = 1] = \alpha/(\alpha + \beta).$$

Thus the equilibrium distribution function for the *SMD*, X, for an element of basin storage with depth s is

$$F(x, s) = \{\alpha F^W(x, s) + \beta F^D(x, s)\}/(\alpha + \beta)$$

where $F^D(x, s)$, $F^W(x, s)$ are the solutions of (5.14) and (5.15) when the suffices $i, i + 1$ are omitted. Thus the basin-wide equilibrium distribution function for *SMD* is

$$(\alpha + \beta)^{-1} \int_{s=0}^{\infty} \{\alpha F^W(x, s) + \beta F^D(x, s)\} p(s; \theta) \mathrm{d}s. \tag{5.17}$$

5.5 THE DISTRIBUTION FUNCTION OF GENERATED RUNOFF

The approach used in section 5.4 can be extended to obtain the distribution function of the runoff generated when elements of basin storage of depth s become filled and begin to spill. Denoting by R_i the runoff generated in the ith time interval Δt, we have for $x \geq 0$ that the conditional distribution function of R_i

$$\begin{aligned} P[R_i \leq x \mid J_i = 1] = {} & P[R_i \leq x \mid J_{i-1} = 0, J_i = 1].P[J_{i-1} = 0 \mid J_i = 1] \\ & + P[R_i \leq x \mid J_{i-1} = 1, J_i = 1].P[J_{i-1} = 1 \mid J_i = 1] \end{aligned}$$

by the theorem of total probability, so that

$$P[R_i \leq x \mid J_i = 1] = \beta \int W(x + z) \mathrm{d}F_{i-1}^D(z) + (1 - \beta) \int W(x + z) \mathrm{d}F_{i-1}^W(z).$$

Integrating by parts, and setting $F_{i-1}^D(0) = 0, F_{i-1}^D(s) = 1$, we obtain

$$P[R_i \le x \mid J_i = 1] = \int_{z=0}^{s} [\beta\{1 - F^D_{i-1}(z)\}$$

$$+ (1 - \beta)\{1 - F^W_{i-1}(z)\}]w(x + z)\mathrm{d}z + W(x). \quad (5.18)$$

This distribution function is conditional on the event $J_i = 1$, namely that the ith time interval Δt is wet; since no additional runoff is generated when $J_i = 0$, when the ith interval is dry, we must have

$$P[R_i \le x \mid J_i = 0] = 1 \quad \text{for all } x \ge 0.$$

We can also derive an "equilibrium" form for the distribution function of generated runoff, pertaining to the case where the effects of initial conditions (at time $i = 0$) no longer have influence. For this, $P[J_i = 0]$ is set equal to $\beta/(\alpha + \beta)$ for large i, and we write $F_{i-1} \to F$ and $R_i \to R$. Then the equilibrium distribution function for generated runoff is

$$P[R \le x] = \beta/(\alpha + \beta) + \alpha/(\alpha + \beta)$$

$$\times \left(\int_{z=0}^{s} [\beta\{1 - F^D(z)\} + (1 - \beta)\{1 - F^W(z)\}]w(x + z)\mathrm{d}z \right.$$

$$\left. + \mathrm{W(x)} \right). \quad (5.19)$$

As before, this expression refers to the runoff generated from an element of storage of depth s, occurring at density $p(s, \theta)$ in the basin. To obtain the distribution function of mean depth of runoff generated from the entire basin, it is necessary to multiply (5.19) by $p(s; \theta)$ and integrate with respect to s between the limits 0 and ∞.

5.6 THE DISTRIBUTION FUNCTION OF AREA CONTRIBUTING TO SURFACE RUNOFF

Assuming the equilibrium conditions of the preceding section ($P[J_i = 0] = \beta/(\alpha + \beta)$ for large i; $F_{i-1} \to F$; $R_i \to R$) it is possible to derive an expression for distribution function of the area within a drainage basin that is contributing to surface runoff at the end of any time interval during a sequence of such intervals in which rain falls. For an element of area with storage S lying in the interval $(s, s + \mathrm{d}s)$, the probability $P[R = 0]$ is the probability that this element produces no surface runoff, and from (5.19) we have

$$P[R = 0] = \alpha/(\alpha + \beta).$$

$$\left(\int_{z=0}^{s} [\beta\{1 - F^{D}(z)\} + (1 - \beta)\{1 - F^{W}(z)\}]w(z)dz \right) \tag{5.20}$$

But elements of area with small storage depth S will fill more rapidly, and hence surface runoff will occur sooner, than those with large S, so if Π is the proportion of basin area for which $S \leq p^*$, i.e.

$$\Pi = \int_{s=0}^{p^*} p(s)ds \tag{5.21}$$

then we have

$$P[\text{proportion of area contributing to surface runoff is} \leq p^*]$$

$$\alpha/(\alpha + \beta). \int_{s=0}^{\Pi^{-1}(p^*)} [\beta\{1 - F^{D}(z)\} + (1 - \beta)\{1 - F^{W}(z)\}]w(z)dz,$$

where $\Pi^{-1}(p^*)$ is the value of Π satisfying (5.21) for p^* given. Recalling that $F^{D}(.)$, $F^{W}(.)$ depend on s and should strictly be written $F^{D}(.;s)$, $F^{W}(.;s)$, it would be more correct to write

$$P[\text{proportion of area contributing to surface runoff is} \leq p^*]$$

$$\alpha/(\alpha + \beta). \int_{s=0}^{\Pi^{-1}(p^*)} [\beta\{1 - F^{D}(z;p^*)\} + (1 - \beta)\{1 - F^{W}(z;p^*)\}]w(z)dz.$$

The result illustrates the difference between the distribution function model and the "single store, single threshold" rainfall–runoff model of widespread usage; for the latter, the proportion of contributing area is zero if the threshold is not exceeded, and one otherwise.

5.7 FREQUENCY RELATIONSHIPS BETWEEN GENERATED RUNOFF AND RAINFALL

Under equilibrium conditions,

$$P[W > u] = \{1 - W(u)\}\alpha/(\alpha + \beta)$$

so that the return period of the rainfall u is

$$T = 1/[\{1 - W(u)\}/\alpha/(\alpha + \beta)]$$

time intervals. The rainfall of return period $T.\Delta t$ is therefore

$$u_T = W^{-1}(1 - (\alpha + \beta)/\alpha T). \tag{5.22}$$

But since

$$P[R \leq x \mid W = u] = \int_{s=0}^{\infty} [\beta\{1 - F^D(u - x)\} + (1 - \beta)\{1 - F^W(u - x)]p(s)ds$$

the median runoff resulting from the rainfall u_T is the runoff r_T which satisfies

$$\int_{s=0} [\beta\{1 - F^D(u - x)\} + (1 - \beta)\{1 - F^W(u - x)]p(s)ds = \tfrac{1}{2}. \tag{5.23}$$

Hosking and Clarke (1989, 1990) show that the return period of this runoff r_T is

$$\begin{aligned} 1/\{1 - P[R \leq r_T]\} = (\alpha + \beta)/\alpha.\{1 - W(r_T) \\ - \int_{s=0}^{\infty}\int_{z=0}^{s} [\beta\{1 - F^D(z)\} + (1 - \beta) \\ \times \{1 - F^W(z)]w(z + r_T))p(s)dzds\}^{-1} \end{aligned} \tag{5.24}$$

Thus when (1) rainfall in successive time intervals of length Δt is represented by a two-state, first-order Markov chain with transition probabilities determined by α, β; (2) rainfall in wet periods, and potential evaporation in dry periods, are serially independent with distribution functions $W(x)$ and $D(x)$ respectively, with pdfs $w(x)$ and $d(x)$, then the rainfall u_T having return period $T.\Delta t$ is given by (5.22) in terms of α, β and $W(x)$. The runoff generated by this rainfall u_T has a distribution function with median r_T, the solution of (5.23), and the return period of this generated runoff is given by (5.24).

The limitation of this result is that it refers to the surface runoff generated by the rainfall event with return period $T.\Delta t$, and not to the runoff appearing at the catchment outfall: in other words, no account has yet been taken of the routing process by which generated runoff becomes observed flow. Hosking and Clarke (1989, 1990) extend their analysis to the case where the generated runoff is routed through a linear reservoir: if this reservoir has contents U_i at the end of the ith time period, the discharge from it is taken to be $Q_i = \lambda(U_{i-1} + R_i)$, where R_i is the runoff generated during period i. After Q_i has been removed from the reservoir, its contents are then

$$U_i = (1 - \lambda).(U_{i-1} + R_i).$$

Hosking and Clarke (1989, 1990) derive an expression for the distribution function $P[Q_i \leq q]$ of the form

$$P[Q_i \leq q] = G_i(s, q(1 - \lambda)/\lambda)$$

giving forms for the G_i, but the expressions are complicated and likely to be of limited application, even for the simple structure (linear reservoir) assumed for the routing procedure. The conclusion is that it is by no means easy to derive theoretical results for rainfall–runoff models of general form with storage distributed across a river basin according to the density function $p(s, \theta)$. Nevertheless, there appears to be considerable scope for determining both the form of this density function, for estimating its parameters, and indeed for extending the model structure to include several layers of storage and more complex routeing processes, by Monte Carlo simulation.

5.8 FURTHER CONSIDERATIONS

First, the development given above expressed the distribution functions $F_i^W(x)$ and $F_i^D(x)$ of soil moisture deficit at the ends of wet and dry time intervals respectively, in terms of the distribution functions $W(x)$ and $D(x)$ of precipitation and potential evaporation: and $W(x)$ and $D(x)$ were assumed not to vary in time. But the expressions for $F_i^W(x)$ and $F_i^D(x)$ will not change if $W(x)$ and $D(x)$ are replaced by time-dependent distribution functions $W_i(x)$ and $D_i(x)$, thereby allowing for seasonal dependence in precipitation and potential evaporation.

Second, the formulation of rainfall–runoff models given above in terms of general storage functions $p(s, \theta)$ includes the family of "lumped" rainfall–runoff models as a special case, since a "lumped" storage is equivalent to the particular form of $p(s, \theta)$ in which $p(s, \theta)$ has unit probability mass for some unknown storage s_0, and zero probability for $s \neq s_0$.

Moore (1985) presents some results for two important developments of the distribution function model described here. He considers the following cases:

1. Where the model is extended to allow for drainage to aquifer storage supplying baseflow. By considering that a storage element is open at the bottom allowing drainage to occur at a constant rate γ until the storage element becomes empty, he shows how to calculate the instantaneous drainage rate $b(t)$ of the flux to groundwater.
2. Where the assumption is relaxed that storage elements act independently. Moore (1985) considers two options, the more acceptable of which is as follows. Water is redistributed within the basin in such a way that a constant deficit, say, is maintained across the population of stores, except for those stores having depth s less than $D^*(t)$ which will be empty.

Moore (1985) gives a full discussion of these cases, including a modification to (2) where there is also deep drainage to aquifer storage.

5.9 EXTENSION TO THE MODELLING OF BASIN SEDIMENT YIELD

Whilst the transport of sediment in river channels has received much attention (Raudkivi, 1967; Graf, 1971; Bettes and White, 1981; Soni, 1981) there has been rather less emphasis on the more challenging problem of modelling processes operating at the basin scale, where sediment becomes available for removal over land surfaces, and is then removed by water and tranported to the stream channels. Negev (1967) developed a basin sediment yield model which incorporated some of the basic hydrological principles (variability through time in the quantity of sediment available within the basin for removal; capacity of the basin to transport sediment to the basin outlet); however, shortcomings of the Negev model are its many parameters, and its inclusion of non-differentiable functional relationships which complicate optimisation procedures used to estimate parameters from data.

To avoid such difficulties, Moore and Clarke (1983) sought to extend the kind of ideas used in earlier sections of this chapter to the modelling of basin sediment availability and transport through the drainage basin. The starting-point of the model was the following proposition: after an element of basin area has ceased to contribute to runoff (say at time t_0) sediment that is available for future removal begins to be generated at a maximum rate R_0, and thereafter at a rate defined by two-parameter exponentially decreasing curve of the form

$$R(t) = R_0 \exp[-K(t - t_0)] \qquad (t > t_0),$$

where K is a decay-rate parameter. At subsequent time t, the depth of sediment available for removal is then

$$d(t) = R_0 \int_{\tau = t_0}^{t} \exp[-K(\tau - t_0)]\mathrm{d}\tau$$

$$= R_0 K^{-1}[1 - \exp[-K(t - t_0)].$$

In the absence of rainfall, sediment continues to accumulate at an exponentially decreasing rate, asymptotically approaching a maximum depth $d(\infty) = R_0/K$; this depth then remains on the element of basin area, until such time as that element contributes to basin runoff, when the sediment is completely removed; thereafter, sediment would be regenerated at the instantaneous rate $R(t)$.

5.10 CONCLUSION

As is often the case, the realistic representation, in probabilistic terms, of the physical processes by which rainfall is transformed into runoff and evaporation lost to the atmosphere, is far from straightforward. Analytical results can be obtained for relatively simple cases (storage elements acting independently; no account taken of relative spatial configuration of storage elements), but more complex cases must be treated by simulation. Nevertheless, with the ever-increasing power of desk-top computers, simulating the behaviour of distribution function models when the storage density $p(s, \theta)$ is allowed to vary in form, and when the parameters θ are allowed to vary, becomes increasingly possible. The value of analytical results, even of much-simplified cases, remains, in that they provide insight into what might be expected from simulation, and a check that simulations when executed are correct.

Chapter Six

The Poisson and related processes

6.1 THE IMPORTANCE OF THE POISSON PROCESS

The Poisson process forms the starting-point for very many applications in the water sciences in which probabilistic models are used to describe events occurring irregularly in time or space. Examples include: movement of sediment particles; development of storm cells; occurrence in a river of peak discharges greater than some threshold value; arrival times of raindrops on leaf surfaces. Even where physical or biological processes are too complex to be adequately represented by a Poisson process, it often serves as a starting-point for more sophisticated models; and where analytical descriptions of the processes are not possible, the Poisson process serves as a "yardstick" with which the results of computer simulations may be compared. The following section reviews some basic properties of the Poisson process; section 6.3 presents some generalisations; and the remainder of the chapter presents some applications.

6.2 REVIEW

In Chapter Two it was shown that the Poisson process described phenomena for which the probability of an event occurring in any small time interval Δt is proportional to Δt, such that events occur independently in any non-overlapping short time intervals Δt_1 and Δt_2. More precisely, for the random variable X to occur in $(t, t + \Delta t]$,

(i) $P[X = 1] = \lambda.\Delta t + o(\Delta t)$
(ii) $P[X = 0] = (1 - \lambda.\Delta t) + o(\Delta t)$
(iii) $P[X > 1] = o(\Delta t)$.

If Y denotes the number of occurrences in a period $(0, T]$, the pdf of Y can be obtained in various ways, one of which is the following. Suppose that $P_y(t)$ is the probability that y events have occurred in the interval $(0, t]$; then, since the occurrence of any event in $(t, t + \Delta t]$ is by definition independent of occurrences in $(0, t]$,

$$P_y(t + \Delta t) = P_{y-1}(t).\lambda\Delta t + P_y(t).(1 - \lambda\Delta t) + o(\Delta t)$$

which after some rearrangement gives

$$\{P_y(t + \Delta t) - P_y(t)\}/\Delta t = \lambda P_{y-1}(t) - \lambda P_y(t)$$

and as Δt tends to zero,

$$\mathrm{d}P_y(t)/\mathrm{d}t = \lambda\{P_{y-1}(t) - P_y(t)\}.$$

The method of probability generating functions can be used to solve this differential equation. Multiplying by z^y, adding over y from $y = 0$ to infinity, and using the fact that $P_{-1}(t)$ is zero (since the random variable Y is restricted to non-negative values) we have

$$\mathrm{d}Q(z)/\mathrm{d}t = \lambda(z - 1)Q(z)$$

where Q denotes the probability generating function of Y. This equations does not involve y, and can be integrated to give

$$Q(z) = \exp\{\lambda(z - 1)t\} \tag{6.1}$$

using the initial condition $Q = 0$ for $t = 0$. Recalling from Chapter Two that the coefficient of z^y in the expansion of $Q(z)$ gives the probability $P[Y = y]$, we obtain

$$P[Y = y] = (\lambda t)^y \exp(-\lambda t)/y!, \tag{6.2}$$

a result that can also be found by (for example) considering the limit, as N tends to infinity, of the binomial

$$P[Y = y] = N!/\{y!(N - y)!\}(\lambda\Delta t)^y(1 - \lambda\Delta t)^{N-y},$$

where $\Delta t = t/N$.

Chapter Two also stated that time intervals between events have an exponential distribution. If the random variable T, with distribution function $F_T(t) = P[T \leq t]$, represents the time to the first event measured from $t = 0$, we have from (6.2) $P[Y = 0] = \exp(-\lambda t)$, so that

$$P[T > t] = \exp(-\lambda t)$$

giving

$$1 - P[T \leq t] = \exp(-\lambda t) \quad \text{and} \quad F_T(t) = 1 - \exp(-\lambda t).$$

Differentiation of the distribution function gives the pdf of T, namely

$$f(t) = \lambda\exp(-\lambda t), \tag{6.3}$$

and evaluation of $E[T]$, $E[T^2] - E[T]^2$ shows that the pdf $f(t)$ has mean $1/\lambda$ and variance $1/\lambda^2$.

6.3 SOME GENERALISATIONS

6.3.1 Spatial processes

In section 6.2, events were taken to occur in time, with the random variables Y and T denoting the number of events occurring in a time interval of length t, and the interval between events respectively. Equally, Y and T could denote the number of events occurring in a one-dimensional spatial interval of length s, and the distance between events; an example might be the occurrence of plants of a given species distributed along a river channel. More generally still, Y can represent the number of events occurring in a two- (three-) dimensional spatial region of area (volume) S; the random variable T then generalises to the distance between an event and its nearest neighbour. In Chapter Two, the interval T_r giving the sum of the time intervals

$$X_1, X_2, \ldots X_r \qquad (T_r = X_1 + X_2 + \ldots + X_r)$$

was given as a particular case of the gamma distribution in which the parameter κ is an integer, $\kappa = r$; where the Poisson process is generalised to two or three dimensions, the analogue of T_r becomes the distance from an event to its rth nearest neighbour. For the case $r = 1$, the pdf of the distance D from a point to its nearest neighbour can be shown to be (Cox and Isham, 1980)

$$f(d) = 2\lambda\pi d \exp(-\lambda\pi d^2)$$

whilst for $r = k$, the pdf of the distance D to its kth nearest neighbour is

$$f(d) = 2(\lambda\pi)^{k-1} d^{2k-1} \exp(-\lambda\pi d^2)/(k-1)!.$$

6.3.2 Poisson process with time-dependent $\lambda \equiv \lambda(t)$

In section 6.2, it was assumed that the Poisson parameter or "rate constant" λ did not vary with time. However, suppose in the equation

$$\mathrm{d}Q(z)/\mathrm{d}t = \lambda(z-1)Q(z)$$

of section 6.2, that the parameter λ is time-dependent, $\lambda \equiv \lambda(t)$, giving

$$\mathrm{d}Q(z)/\mathrm{d}t = \lambda(t).(z-1)Q(z).$$

Integration gives the form

$$Q(z) = \exp\{(z-1)\Lambda(t)\},$$

where

$$\Lambda(t) = \int_{\infty}^{u=t} \lambda(u)\mathrm{d}u. \tag{6.4}$$

As before, the coefficient of z^y gives the probability $P[Y = y]$:

$$P[Y = y] = \{\Lambda(t)\}^y \exp(-\Lambda(t)\}/y!$$

showing that when λ is a function of time, the process is still Poisson, but with parameter given by (6.4).

Models of this kind may be considered, for example, where suspended sediment particles moving along a channel are brought into suspension at a varying rate $\lambda(t)$, where time dependence is a consequence of time-varying discharge $q(t)$.

6.3.2 Combinations of several Poisson processes

Suppose that k separate Poisson processes are occurring independently: for example, suppose that sediment particles of relatively indistinguishable sizes have densities having k distinct values. The densities may, for example, be derived by approximating a continuous distribution of densities by a histogram with k classes. At a fixed measuring point in a channel, the times are recorded at which particles pass the measuring point, and the k Poisson parameters are denoted by $\lambda_1, \lambda_2, \lambda_3 \ldots \lambda_k$. If

$$N(t, t + \Delta t), N^{(i)}(t, t + \Delta t)$$

denote the numbers of all particles of whatever type, and the numbers of particles of type $i(i = 1, 2, \ldots k)$ respectively, passing in the interval $(t, t + \Delta t]$, then

$$P[N(t, t + \Delta t) = 0] = \prod_{i=1}^{k} P[N^{(i)}(t, t + \Delta t) = 0]$$

$$= \Pi(1 - \lambda_i \Delta t)$$

which, ignoring terms containing powers of Δt higher than the first, is

$$1 - \lambda \Delta t, \tag{6.5}$$

$$\text{where } \lambda = \lambda_1 + \lambda_2 + \lambda_3 + \ldots + \lambda_k.$$

Also

$$P[N(t, t + \Delta t) = 1] = \sum_{i=1}^{k} P[(N^{(i)}(t, t + \Delta t) = 1)$$

$$\cap (N^{(j)}(t, t + \Delta t) = 0)]$$

where the summation is over $i \neq j$;

$$= \lambda \Delta t. \tag{6.6}$$

From (6.5) and (6.6), we see that the particles with the k densities move according to a Poisson process with parameter

$$\lambda = \lambda_1 + \lambda_2 + \lambda_3 + \ldots + \lambda_k.$$

In practice, the simple Poisson model is likely to be inadequate because of factors such as coalescence and decomposition of particles which render invalid the assumption of independence of the k processes.

6.3.3 Multiple occurrences

Chapter Two gave three examples (Neyman type A, Pólya–Aeppli, Poisson–Pascal) of pdfs obtained when events occur in clusters, the clusters having a Poisson distribution and the number of events per cluster having respectively a geometric, Poisson and negative binomial distribution. Suppose, more generally, that (1) the number of clusters has a Poisson distribution with parameter λ; (2) the number of events per cluster is a random variable C with pdf

$$P[C = i] = \phi_i \qquad (i = 0, 1, 2, \ldots)$$

with probability generating function

$$\Lambda(z) = \phi_0 + \phi_1 z + \phi_2 z^2 + \phi_3 z^3 + \ldots.$$

Then from the material of section 2.2 it can be shown that the probability generating function of N, the number of events occurring in the interval $(0, t]$, is

$$\exp\{\lambda t \Lambda(z) - \lambda t\}$$

or

$$\exp(-\lambda t).\exp\{\lambda t \Lambda(z)\}. \tag{6.7}$$

By expanding the second factor $\exp\{\lambda t \Lambda(z)\}$ as a power series in z, the pdf $P[N = n]$ can be found, and differentiation of (6.7) yields its moments as shown in section 2.2. Models of this general form may provide a first approximation to suspended particles which coalesce, perhaps with time-varying rate of coalescence.

6.4 SOME APPLICATIONS

This section presents some particular applications of Poisson processes from the fields of flood frequency studies, rainfall interception modelling, and

Table 6.1 Numbers of peak daily flows $q_t(q_{t-1} < q_t > q_{t+1})$ exceeding $100\,\mathrm{m}^3\,\mathrm{s}^{-1}$, River Itajaí-Açú, southern Brazil: 1935–44. Months 1 to 12 are January to December

	Month											
	1	2	3	4	5	6	7	8	9	10	11	12
1935	2	2	4	2	1	0	0	3	1	4	4	2
1936	4	2	2	3	2	2	3	1	1	1	0	0
1937	2	3	3	3	4	3	0	0	3	3	3	4
1938	3	5	1	3	1	0	1	4	0	2	2	4
1939	3	7	4	1	4	4	1	2	3	3	4	4
1940	2	2	5	2	2	3	2	2	3	1	0	1
1941	1	3	0	0	1	3	3	2	3	5	1	1
1942	4	0	2	1	0	0	0	1	0	0	2	0
1943	0	3	0	1	0	0	1	0	1	2	0	1
1944	3	5	4	1	2	3	3	1	0	3	2	2

hydrograph simulation. The examples given are one-dimensional, discussion of two-dimensional models involving Poisson events is given in the next chapter.

6.4.1 Flood peaks over a threshold

Table 6.1 shows the numbers of peak mean daily discharges q_t exceeding $100\,\mathrm{m}^3\,\mathrm{s}^{-1}$ (values such that $q_{t-1} < q_t > q_{t+1}$) in each month of the 10-year period 1935–44, for the River Itajaí-Açú in southern Brazil. In an analysis of such data, we might wish to examine whether the frequency of such peaks has changed with time, possibly as a consequence of land-use change; there is evidence that deforestation results in more rapid runoff and reduced infiltration (Bruijnzeel, 1996). Initially, we consider whether peaks can be considered to be a Poisson process, and if so whether there is evidence that the Poisson parameter varies with time. The analysis ignores differences in month lengths and leap years.

Ignoring time dependence for the moment, consider whether the 120 data values look like observations of a Poisson variable. Their mean is 1.99, their variance 2.43, so there is some evidence of "overdispersion" with variance greater than the mean. Fitting a Poisson by maximum likelihood gives $\hat{\lambda} = 1.99 \pm 0.129$, with observed frequencies, and those expected on the Poisson hypothesis, as follows:

x	0	1	2	3	4	5 +
Observed	25	24	25	26	15	5
Expected	16.33	32.57	32.48	21.59	10.76	6.26

and a deviance of 10.91 on 4 df. Using an approximate χ^2 test, the tabulated value for the 5% significance level and 4 df is 9.488, so that the deviance is rather larger than would be expected on a Poisson hypothesis. However, it is possible that the departure may be a consequence of time variation in the

Poisson parameter. Possible explanatory variables, bearing in mind that the data are monthly observations, are

$$c1 = \cos(2\pi t/12), s1 = \sin(2\pi t/12),$$

possibly also including $c2 = \cos(4\pi t/12)\ldots$, where $t = 0, 1, \ldots 119$. Using a log link function, fitting the model

$$\log_e \lambda = \beta_0 + \beta_1 c1 \tag{6.8}$$

gives as the estimate of β_0, 0.6728 ± 0.0657, and as the estimate of β_1, 0.2550 ± 0.0926, so that the estimate of β_1 is 2.75 times its standard error and suggests a significant seasonality in the Poisson parameter λ. Use of $c1$ as an explanatory variable reduces the total deviance from 169.3 on 119 df to 161.7 on 118 df, the difference 7.7 having 1 df. Inclusion of $s1$ as an additional explanatory variable results in only a minor reduction in deviance of 0.581 with 1 df; similarly for the inclusion of $c2$ and $s2$.

To look at whether there is evidence of significant differences between years, the model (6.8) was modified to include a "year" component, y_i, so that the model became

$$\log_e \lambda = \beta_0 + \beta_1 c1 + y.$$

The constant y was fitted for each year, resulting in a further reduction in deviance of 35.2 on 9 df amongst the 10 years; so there do appear to be year effects in addition to the seasonal variation in λ. The year effects which were large relative to their standard errors were those for years 1942 and 1943, for which the year effects were estimated as -0.916 ± 0.374 and -1.022 ± 0.389 respectively; inspection of Table 6.1 suggests that peaks in this years behaved differently.

With a longer record of peak occurrences, it would perhaps be worth looking for a time trend in the year constants y, which might be indicative of a significant land-use effect.

6.4.2 Distribution of annual maximum flood

Suppose that a continuous hydrograph recorded at a flow-gauging station is processed to produce, for each complete year of record, (1) the number of peak discharges exceeding a threshold value, q_0 say; (2) the instantaneous peak discharge associated with each such exceedence. In the statistical terminology, the resulting data is from a **marked point process**: with each event (in this case, the occurrence of a peak) there is associated another random variable (possibly more than one) termed the mark. In the present case, there is just one mark, namely the magnitude of q, the discharge where the peak occurs.

Now suppose that the number of peak discharges in a year (including those that are less than the threshold q_0) is a Poisson random variable N with parameter λ; suppose also that π is the conditional probability

$$\pi = P[Q \geq q \mid q \geq q_0] \tag{6.9}$$

that the peak discharge for any event exceeds some $q \geq q_0$ (i.e. conditional on it exceeding the threshold q_0). Then, when n, the number of peaks in the year is given, the probability that r of them also exceed a value $q \geq q_0$ is, using the binomial distribution,

$$P[R = r \mid N = n] = n!/[r!(n - r)!].\pi^r.\{1 - \pi\}^{n-r}.$$

But the number of peaks n is a Poisson variable, so that the unconditional probability $P[R = r]$ is, by the theorem of total probability given in Chapter Two,

$$\begin{aligned} P[R = r] &= \sum_{n=r}^{\infty} P[R = r \mid N = n].P[N = n] \\ &= \sum_{n=r}^{\infty} n!/[r!(n - r)!].\pi^r.\{1 - \pi\}^{n-r}.\exp(-\lambda).\lambda^n/n! \\ &= \{\exp(-\lambda).(\lambda\pi)^r/r!\}.\sum_{n=r}^{\infty}(n - r)!\{1 - \pi\}^{n-r}\lambda^{n-r} \\ &= \{\exp(-\lambda\pi).(\lambda\pi)^r/r!\}. \end{aligned} \tag{6.10}$$

Now consider the distribution of the annual maximum discharge, $Q_{\max}$ say, with distribution function $P[Q_{\max} \leq q]$. But if the annual maximum is less than q then R, the number of peaks exceeding q in (6.10), must be zero; hence

$$P[Q_{\max} \leq q] = \exp(-\lambda\pi). \tag{6.11}$$

In the *NERC Flood Studies Report* (1975), the magnitudes of the marked variable were assumed to be exponentially distributed: that is,

$$\pi = P[Q \geq q \mid q \geq q_0] = \exp\{-(q - q_0)/\beta\} \tag{6.12}$$

and the distribution function of $Q_{\max}$ then becomes

$$P[Q_{\max} \leq q] = \exp(-\lambda.\exp\{-(q - q_0)/\beta\}) \tag{6.13}$$

which is the distribution function of the Gumbel distribution widely used in flood frequency analysis. As with the example of section 6.4.1, it is straightforward to extend this basic model to allow for time-varying λ; a log-link function is suggested, and again seasonality in the parameter value would be introduced by a linear predictor of the form

$$\eta_t = \beta_0 + \beta_1\cos(2\pi t/12) + \beta_2\sin(2\pi t/12) + \text{higher terms if required.}$$

The *NERC Flood Studies Report* (1975) gives a carefully argued discussion concerning the Poisson assumption for rivers in the United Kingdom, and the nature of the departure from Poisson behaviour. Analysis showed that some years had a very large number of peaks and that peaks over the threshold q_0

sometimes occurred in clusters. The magnitudes of successive peaks did not appear to be statistically dependent, but the degree of clustering was such that times between peaks were not identically distributed, nor were they always independent; independence was a reasonable assumption in summer months, but was less reasonable in winter months. As an alternative to the Poisson assumption, the negative binomial distribution was used for $P[N = n]$ in expression (6.10) above (recall that the negative binomial is obtained, amongst other options, where clusters have a Poisson distribution, whilst the number of peaks within clusters has a gamma distribution). No formal tests were made, but the two-parameter negative binomial appeared to give little improvement over the one-parameter Poisson distribution.

6.4.3 Interception of raindrops by leaf surfaces

Calder (1986) has proposed a simple probabilistic model of rainfall interception by a plant canopy, subsequently extended in a series of papers (Calder, 1996; Calder et al., 1996). The 1986 model did not consider the rates of arrival of raindrops as a stochastic process developing in time: the probability of a randomly chosen element of canopy area (e.g. leaf) being struck by x raindrops was taken as Poisson such that

$$P[X = x] = \exp(-\lambda)\lambda^x/x!.$$

Clearly, the probability of a leaf being struck by r raindrops or fewer is then

$$S = \sum_{x=0}^{r} P[X = x]$$

and by more than r raindrops $1 - S$. If each leaf can retain a maximum of q raindrops, with any excess shedded, the mean number of drops retained per leaf is

$$n = (1 - S)q + \sum_{x=0}^{r} x \exp(-\lambda)\lambda^x/x!$$

or

$$n = \exp(-\lambda) \sum_{x=0}^{r} (x - q)\lambda^x/x!.$$

If L is the number of leaves per unit ground area and v the mean volume of raindrops, the depth of storage on the canopy per unit ground area is $C = nvL$, the maximum depth of canopy storage is $C_{max} = qvL$ and the mean number of raindrop strikes is related to the rainfall P recorded at ground level is $m = P/vL$.

The basic model could be modified in a number of ways. If raindrops (still considered to be of equal volume v) are considered to arrive according to a

Poisson process with parameter μ with time t measured from the start of a rainstorm when the interception store is empty, the depth per unit leaf area in interception storage at any subsequent time t, when r raindrops have been intercepted, will be rv when $r < C_{max}/v$, and C_{max} otherwise; so that

$$P[C = rv; t] = \exp(-\mu t)(\mu t)^r/r! \qquad (r < C_{max}/v)$$

$$P[C = C_{max}; t] = \exp(-\mu t) \sum_{r=C_{max}/v}^{\infty} (\mu t)^r/r! \qquad (r \geq C_{max}/v)$$

whence the mean and variance of interception storage can be calculated as a function of time. It is also straightforward to introduce time variation in the parameter μ, corresponding to varying rainfall intensity.

In a later paper (Hall et al., 1996), the effects of drop-size distribution were studied, the authors citing a relation of the form

$$N_A(d) = v(d)N_0\exp\{-4.1R^{-0.21}d\}$$

relating the number N_A(d) of drops with diameter d to the terminal velocity $v(d)$ and rainfall intensity R. For given R, this relation can be used to derive a pdf for raindrop volume (assuming spherical drops) in terms of a discrete element of raindrop volume δv; if the probability generating function of raindrop volume is then

$$\Lambda(z), \text{ i.e. } \Lambda(z) = \phi_0 + \phi_1 z + \phi_2 z^2 + \phi_3 z^3 + \ldots$$

where ϕ_j is the probability of a raindrop with volume $j.\delta v$, the model then becomes of the form (6.7) above.

6.4.4 Modelling bedload transport in upland trout streams

Hurley (1992) used a model more general than the Poisson process (but including it as a special case) for the interpretation of bedload movement in two streams (Great Eggleshope Beck, Carl Beck) in Upper Teesdale, England. Sediment traps were installed in the each stream-bed; a bedload event was defined as any discharge which transported coarse sand and granules greater than 2 mm into a trap; 78 such events were recorded in Great Eggleshope and 82 in Carl Beck during a six-year period from December 1978 to 6 June 1984. The data showed a seasonal cycle with bedload events occurring more frequently in winter than in summer, together with a possible small spring peak.

The essence of Hurley's approach is (1) the modelling of bedload event occurrences using an inhomogeneous (time-dependent) gamma process, with the inhomogeneous Poisson process as a special case; (2) the relation of the intensity $\lambda(t)$ to harmonic terms involving

$$\cos(\omega t), \sin(\omega t), \cos(2\omega t), \sin(2\omega t)\ldots(\omega = 2\pi/365.25),$$

and to a simple time trend, using a generalised linear model (GLM) with log-link function of the form

$$\log_e \lambda(t) = \beta_0 + \beta_1 t + \beta_2\cos(\omega t) + \beta_3\sin(\omega t) + \beta_4\cos(2\omega t) + \beta^5\sin(2\omega t) + \beta_6\cos(3\omega t) + \beta_7\sin(3\omega t). \quad (6.14)$$

The data consisted of the $m + 1$ times $t_0, t_1, t_2, \ldots t_m$ on which bedload events occurred, and t_0 was taken as the date of the first event. Analogous to section 6.3.2, the inhomogeneous gamma process describes the pdf of t_j, conditional on t_{j-1}, as

$$f(t_j | t_{j-1}) = \lambda(t_j)[\Lambda(t_j) - \Lambda(t_{j-1})]^{\kappa-1}\exp\{-[\Lambda(t_j) - \Lambda(t_{j-1})]\}/\Gamma(\kappa)$$

where

$$\Lambda(t) = \int_{t_0}^{t} \lambda(u)\mathrm{d}u.$$

The transformation to $\Lambda(t)$ can be regarded as a transformation from the original time-scale measure by t to a new time scale in which the time intervals between events $\Lambda(t_j) - \Lambda(t_{j-1})$ have a gamma distribution with unit scale parameter, and shape parameter κ; the particular case $\kappa = 1$ giving the inhomogeneous Poisson process, but Hurley (1992) also considered $\kappa = 2$ and $\kappa = 3$, commenting that cases with non-integral κ have no simple physical interpretation. A series of nested models were fitted with (6.14) as the maximal model; for each model fitted, the maximised likelihood was calculated. The maximised likelihood was always less for $\kappa = 3$ than for $\kappa = 1$ or $\kappa = 2$, showing that models with a high degree of "gamma-ness" were not required; and indeed the Poisson model ($\kappa = 1$) gave a maximised likelihood greater than models for $\kappa = 2$ in half the models fitted, whilst for the remainder it was only slightly worse (although the difference was statistically significant). The models with $\kappa = 2$ fitted better where sine and cosine terms in $2\omega t$ were included; terms with sines and cosines in $3\omega t$ were never significant. It is interesting that (when $\kappa = 2$) a significant negative time trend ($\hat{\beta}_1 = -0.658 \times 10^{-3} \pm 0.102 \times 10^{-3}$) was found (see expression for $\log_e \lambda(t)$ in (6.14)), suggesting a tendency for bedload events to become less frequent over the six-year period.

6.4.5 Shot-noise models for simulating continuous streamflow sequences

Methods of time-series analysis are extensively used to model sequences of streamflow where runoff is accumulated over time intervals of a month or so; see, for example, Hipel and McLeod (1994), McLeod and Hipel (1995). For monthly time intervals, model variants of the general Box–Jenkins form are commonly satisfactory, and can be made sufficiently flexible to account for non-stationarity in seasonal correlational structure. They are less satisfactory, however, for modelling sequences with data recorded at daily time intervals or less, which are characterised by rapid rises, often with substantial superimposed noise, followed by longer periods of relatively noise-free recession, as flow diminishes after rainfall ceases. The basis of models of the Box–Jenkins type is that their random shocks have normal distribution, or can be approximated by the normal distribution after a power-law or logarithmic transformation; assumptions of normality are often more difficult to justify with daily streamflow data. In fact, streamflow models derived from normally distributed random variables are always "time-reversible" in the sense that sequences look broadly similar in statistical terms, when the direction of the time axis is reversed; runoff accumulated over time intervals of a day or less clearly does not have this feature.

The "shot-noise" model, originally proposed by Bernier (1970) and subsequently developed by Weiss (1973), is an alternative approach to the modelling of daily flow sequences. It is a particular case of a family of models termed **filtered Poisson processes**, of the following form. Suppose that the events of a Poisson process occur at times $\ldots, t_{-2}, t_{-1}, t_0, t_1, \ldots$; let $y_{-2}, y_{-1}, y_0, y_1, \ldots$ be a sequence of identically distributed random variables independent of the Poisson process; and let $w(t, y)$ be some function. Then the filtered Poisson process is

$$Z(t) = \sum_{m=N(-\infty)}^{N(+\infty)} w(t - \tau_m, y_m).$$

The shot-noise model is another example of a marked point process in which the starting-point is a Poisson process (possibly non-homogeneous), with parameter λ, representing the occurrence in time of flow peaks. Associated with each peak event are the "marks"; these are exponential functions with random amplitudes and a constant decay parameter. Thus the "shot" associated with the peak event at time t_i is of the form $M_i \exp(-\kappa t)$ with $t > 0$, where theM_i are independent and identically distributed random variables independent of the Poisson process of peak events. Thus the shot-noise model can be expressed as

$$Z(t) = \sum_{i}^{T \le t} M_i \exp\{-\kappa(t - T_i)\} \tag{6.15}$$

Cox and Isham (1980) show that (6.15) may be written in first-order, autoregressive form

$$Z(t + \delta) = \exp(-\kappa t) Z(t) + \varepsilon(t)$$

where $\varepsilon(t)$ is an "innovation" term independent of $Z(t)$; the mean, variance, and correlation structure are given, in terms of the moments of the random shock process M_i, as

$$\begin{aligned} \mathrm{E}[Z(t)] &= \lambda \mathrm{E}[M]/\kappa \\ \mathrm{var}[Z(t)] &= \tfrac{1}{2}\lambda \mathrm{E}[M^2]/\kappa \\ \mathrm{corr}[Z(t), Z(t+\tau)] &= \exp\{-\kappa|\tau|\}. \end{aligned} \tag{6.16}$$

Cox and Isham (1980) also give an expression for the moment-generating function of $Z(t)$.

Applications of the shot-noise process for the modelling of daily runoff sequences have been given by Weiss (1977). His starting-point was a Poisson process with parameter λ; the shocks M were taken to have an exponential distribution $f(M) = (1/\theta)\exp(-M/\theta)$, so that three parameters $\{\lambda, \theta, \kappa\}$ defined the shot-noise process. The distribution of $Z(t)$ was then the gamma distribution

$$f(Z) = (1/\theta)^{\lambda/\kappa} Z^{\lambda/\kappa - 1} \exp(-Z/\theta)/\Gamma(\lambda/\kappa).$$

The shot-noise model defined above is in continuous time, and Weiss modified it to the case where streamflow is given as a mean over a period of length T, typically a day. Denoting the discrete process by Z_t, Weiss showed that

$$\begin{aligned} \mathrm{E}[Z_t] &= \lambda\theta/\kappa \\ \mathrm{var}[Z_t] &= (\lambda\theta/\kappa).2[\kappa - (1 - \exp(-\kappa))]/\kappa^2 \\ \mathrm{corr}[Z_j, Z_{j+s}] &= (1 - \exp(-\kappa))^2/[2\{\kappa - (1 - \exp(-\kappa))] \\ &\quad \exp(-\kappa(s-1)) \qquad (s > 1). \end{aligned} \tag{6.17}$$

The shot-noise model was also extended to allow for for seasonal periodicity in the model parameters $\{\lambda, \theta, \kappa\}$; to deal with the case where streamflow sequences are to be modelled at several cross-correlated sites; and to include components corresponding to a slowly varying baseflow component and a rapidly varying surface flow component.

6.4.6 Annual precipitation derived from observed storm sequences

In a remarkable series of papers, Eagleson (1978a, b, c, d, e, f, g) presented a very general statistical dynamic formulation of the vertical water budget at a land-surface interface. Physically based dynamic and conservation equations expressed the infiltration, exfiltration, transpiration, percolation to groundwater and capillary rise from the water table during rainstorms and interstorm periods in terms of independent variables representing the precipitation, potential evapotranspiration, soil and vegetation properties and water-table elevation. Uncertainty was introduced into these equations through the pdfs of the independent climatic variables. The derived pdfs of surface runoff, evapotranspiration, and groundwater discharge were obtained, and the mean values of these pdfs give a long-term average water balance which, to the first order, defines the annual water yield and water loss in terms of annual precipitation, annual potential evapotranspiration, and physical parameters of soil, vegetation, climate and water table. Eagleson's theory is particularly remarkable for its holistic inclusion of water-balance components and for its strong physical basis; so far as it has limitations, it might be claimed that the theory stops at the point where storm rainfall excess is generated, and does not include the subsequent step in which this rainfall excess at a point is routed to the outfall from a drainage basin. For many important applications (such as water-balance studies in semi-arid zones) this may not be important; however, it precludes the use of the theory for (for example) flood prediction. Whilst the models described in Chapter Five are much more rudimentary than that of Eagleson in terms of their description of the physical behaviour of river basins, they include descriptions of the routing processes. Both approaches are limited to the hydrology of temperate and tropical zones; neither has the ability to represent hydrological behaviour in regions where precipitation falls as snow and accumulates until the onset of melting.

The starting-point in the Eagleson theory at which uncertainties are introduced is the way in which precipitation input is represented. Storms are assumed to arrive according to a simple Poisson process, and the duration of each is assumed to be collapsed to a point so that precipitation is regarded as a point process in which the number N of rainfall events occurring in a period $(0, t]$ has a Poisson distribution with parameter ω:

$$P[N = n] = (\omega t)^n \exp(-\omega t)/n! \qquad n = 0, 1, 2, \ldots$$

and the inter-event times are distributed exponentially with pdf:

$$f(t) = \omega \exp(-\omega t).$$

The total precipitation occurring in the n events is

$$Y = \sum_{j=1}^{n} h_j,$$

where h_j is the precipitation depth in the jth point storm. The storm depths were assumed to be independently and identically distributed, although these restrictions can be relaxed as shown below. For the distribution of h, the gamma distribution was selected on the grounds that it provides a good fit with observations over a range of climatic types, and is analytically tractable. Thus the pdf of h was taken as

$$f(h) = \lambda(\lambda h)^{\kappa - 1} \exp(-\lambda h)/\Gamma(\kappa).$$

Then, conditional on n rainfall events having occurred in the interval $(0, t]$, the pdf of the random variable Y is, from the properties of the gamma distribution,

$$f(y \mid n) = \lambda(\lambda y)^{n\kappa - 1} \exp(-\lambda y)/\Gamma(n\kappa).$$

However, the number of storms n in the period $(0, t]$ is also a random variable, and using the relation (Chapter Two)

$$P[A] = P[A \mid B_1] + P[A \mid B_2] + P[A \mid B_3] + P[A \mid B_4] + \ldots,$$

where $B_i: i = 1, 2, \ldots$ is a set of mutually exclusive and exhaustive events (in this case, the events $n = 0, n = 1, \ldots$) the unconditional pdf of the rainfall total Y in the period $(0, t]$ is

$$f(y; t) = \sum_{n=1}^{\infty} \lambda(\lambda y)^{n\kappa - 1}\exp(-\lambda y)/\Gamma(n\kappa).(\omega t)^n \exp(-\omega t)/n!$$

for $y > 0$; and since it is possible that no rainfall events occur in $(0, t]$, there is a discrete "particle" of probability at $y = 0$ given by

$$f(0; t) = \exp(-\omega t).$$

By introducing the further assumption that the coefficient of variation in the length of the rainy season is small, Eagleson used the cumulative distribution function of standardised annual rainfall P_{A} in the form

$$P[P_{\text{A}}/m < z] = \exp(-\omega m_\tau).\left\{1 + \sum_{n=1}^{\infty} (\omega m_\tau)^n/n!.G(n\kappa, \omega m_\tau \kappa z)\right\},$$

where m is mean annual rainfall, m_τ is the mean length of the rainy season and $G(a, x)$ is the incomplete gamma function defined by

$$G(a, x) = \int_{u=0}^{x} u^{a-1}\exp(-u)\mathrm{d}u/\Gamma(a).$$

The cumulative distribution of standardised annual rainfall is thus expressed in terms of two parameters: the product ωm_τ, the mean number of storms per year, and κ, the shape parameter in the gamma distribution of storm depths. The model was tested using annual and hourly data from both moist (Boston, Massachusetts) and dry (Santa Paula, California) climates; using five years of storm data, the variance of the standardised annual rainfall P_{A}/m was smaller than that obtained by more common hydrological methods which use only annual totals.

Eagleson's model assumed that a simple Poisson process for storm arrivals, but it would be straightforward to replace the Poisson distribution

$$P[N = n] = (\omega t)^n exp(-\omega t)/n! \qquad n = 0, 1, 2, \ldots$$

for the number of storm events by the more general alternative

$$P[N = n] = \{\Lambda(t)\}^n \exp\{-\Lambda(t)\}/n! \qquad n = 0, 1, 2, \ldots$$

It also appears likely that the gamma distribution of storm depths could be replaced by an inhomogeneous gamma process as used by Hurley (1992).

6.5 RENEWAL PROCESSES

For the Poisson process, the time between one event and its successor has an exponential distribution: therefore, an obvious generalisation is a process in which inter-event times have some more general distribution $f(x)$. Such processes are **renewal processes**, and are the subject of a classic monograph by Cox (1962). Some principal features of renewal processes (of which the Poisson is a special case) are the following:

1. The time S_n up to the nth event $X_1 + X_2 + X_3 + \ldots X_n$ is the sum of n independently distributed random variables, and statistical theory shows how its distribution can be calculated for general $f(x)$. When n is "large", the central limit theorem shows that S_n has approximately a normal distribution with mean $n\mu$ and variance $n\sigma^2$, where μ, σ^2 are the mean and variance of the distribution $f(x)$ of inter-event intervals.
2. If N_t is the number of events occurring in the interval $(0, t]$, the distribution of N_t is related to that of S_n because $N_t < n$ if, and only if, $S_n > t$; hence
$$P[N_t < n] = P[S_n > t] = 1 - F_n(t),$$
where $F_n(t)$ is the distribution function of N_t. Hence
$$P[N_t = n] = F_n(t) - F_{n+1}(t),$$
with $F_0(t) = 1$. The example of Hurley (1992) given above is an example of the important special case of a renewal process in which inter-arrival times have a gamma distribution with shape parameter κ an integer.
3. The result of (2) above can be used (Cox and Miller, 1965) to show that N_t is approximately (asymptotically) Normally distributed with mean t/μ and variance $\sigma^2 t/\mu^3$.

An **alternating renewal process** has inter-event times which alternate between two distributions, $f_1(x)$ and $f_2(x)$, say. In a large drainage basin such as the Amazon, an individual water molecule may be regarded as following an alternating renewal process, alternating between the liquid and gaseous phases of water, until it is eventually discharged from the basin to the ocean. A further extension is a **simple semi-Markov process**, as used by Rodriguez-Iturbé and Valdés (1979) as a model of river basin behaviour: see Chapter Four. They regarded a particle of water as moving between a number of states according to a p-state Markov chain with $p \times p$ matrix of transition probabilities, as described in the following section.

6.5.1 Renewal process defined on a Markov chain: the geomorphological structure of hydrological response

In the alternating renewal process, inter-event times alternate between two pdfs $f_1(t)$ and $f_2(t)$. An obvious generalisation has k pdfs $f_1(t), f_2(t), f_3(t) \ldots f_k(t)$ associated with the k states of a Markov chain; more generally still, the time spent in a transition between states i and j, now denoted by $f_{ij}(t)$, may depend on both initial and final states. A model of this kind was the starting point for Rodriguez-Iturbé and Valdes (1979) in their work, already introduced earlier in section 4.2.1, linking the instantaneous unit hydrograph of a river basin (the response in terms of river discharge to unit depth of excess rainfall, instantaneously applied uniformly over its area) to Horton's numbers R_A, R_B and R_L, to an internal scale parameter L_Ω, and to a mean velocity of streamflow v. (Horton's numbers are defined by $R_B = N_\omega/N_{\omega+1}$, where N_ω is the number of streams of order ω; $R_L = L_\omega/L_{\omega-1}$, where L_ω is the mean length of streams of order ω; and $R_A = A_\omega/A_{\omega-1}$, where A_ω is the mean area of basins drained by streams of order ω. R_B, R_L and R_A represent the bifurcation ratio, the length ratio and the area ratio for a drainage system, and their values commonly range between 3 and 5, 1.5 and 3.5, and 3 and 6 respectively.)

Recall from Chapter Four that a Markov chain with discrete states has a matrix of transition probabilities denoted by

$$\mathbf{P} = \begin{bmatrix} p_{11} & p_{12} & p_{13}\cdots & p_{1N} \\ p_{21} & p_{22} & p_{23}\cdots & p_{2N} \\ \cdots & \cdots & & \\ p_{N1} & p_{N2} & p_{N3}\cdots & p_{NN} \end{bmatrix}$$

which in the case of particles of effective rainfall moving through a river basin according to the Rodriguez-Iturbé and Valdes model had the particular form

$$\mathbf{P} = \begin{bmatrix} 0 & p_{12} & p_{13}\cdots & p_{1\Omega} & 0 \\ 0 & 0 & p_{23}\cdots & p_{2\Omega} & 0 \\ \cdots & & \cdots & & \\ 0 & 0 & 0\ldots & 0 & 1 \end{bmatrix},$$

where there are $N = \Omega + 1$ states, Ω being the order of the drainage basin and the extra state is the "bucket" or trapping state.

To simplify the argument to be given below, we assume that $N = 2$ and take the matrix $\mathbf{P}$ of transition probabilities to have the simplified form

$$\mathbf{P} = \begin{bmatrix} p_{11} & p_{12} \\ p_{21} & p_{22} \end{bmatrix}.$$

Thus the simplified form has two states only, and we consider a particle moving between the states 1 and 2.

In the general case where p_{11}, p_{12}, p_{21} and p_{22} are all positive, the particle can move from state 1 to state 2 and conversely, and this is the case developed below; when the particle is a particle of runoff moving from a lower-order basin to a higher-order basin, it will of course be necessary to modify the development given below by setting $p_{ij} = 0$ for $i \leqslant j$, $j < N$ to obtain the Rodriguez-Iturbé-Valdes transition matrix $\mathbf{P}$.

Suppose that the particle was in state i at time $t = 0$. Define $h_{ij}(t)(i,j = 1,2)$, the **renewal density for transitions to state** j, as follows:

$$h_{ij}(t) = \lim_{\Delta t \to 0} P[\text{transition to state } j \text{ occurs in } (t, t + \Delta t) \mid \text{particle in state } i \text{ at } t = 0]/\Delta t.$$

With $f_{ij}(t)$ the pdf of the time intervals between transitions from states i,j, define the function

$$\mathfrak{I}_{ij}(t) = 1 - \int_{u=0}^{t} f_{ij}(u)du = \int_{u=t}^{\infty} f_{ij}(u)du \qquad (i,j = 1,2).$$

Here, $\mathfrak{I}_{ij}(t)$ is the **survivor function**, giving the probability that no transition from state i to state j occurred in $(0, t)$.

Given that the particle was in state i at time $t = 0$, let $\pi_{ij}(t)$ be the probability that the particle is in state j at a later time t. We concentrate first on $\pi_1(t)$, the probability that the particle is in state 1 at time t, given that it was already in that state at time $t = 0$. This could happen in any one of three mutually exclusive ways:

1. The particle could have remained in state 1 for the whole period $(0, t)$, without making any excursion into state 2. The probability of this is
$$p_{11}.\mathfrak{I}_{11}(t).$$
2. The particle may have alternated in some way between states 1 and 2, ultimately entering state 1 at some earlier time $t - u$, after which it remained in that state for a time of duration u. The probability of this alternative is
$$p_{11}. \int_{u=0}^{t} h_{11}(t - u)\mathfrak{I}_{11}(u)du$$
the integral being required since the final entry into state 1 may have occurred at any time in $(0, t)$.
3. Finally, the particle may have alternated in some way between state 1 and 2, ultimately entering state 2 at some earlier time $t - u$, after which it made a transition back to state 1 where it stayed for a time of duration u. The probability of this alternative is
$$p_{21} \int_{u=0}^{t} h_{12}(t - u)\mathfrak{I}_{21}(u)du.$$

These three probabilities being mutually exclusive, their sum must be equal to $\pi_{11}(t)$:

$$\pi_{11}(t) = p_{11}\mathfrak{I}_{11}(t) + p_{11}\int_{u=0}^{t} h_{11}(t-u)\mathfrak{I}_{11}(u)du$$

$$+ p_{21}\int_{u=0}^{t} h_{12}(t-u)\mathfrak{I}_{21}(u)du.$$

Similar equations can be written for

$$\pi_{12}(t), \pi_{21}(t), \pi_{21}(t) \quad \text{and} \quad \pi_{22}(t).$$

When there N states instead of merely two, these equations can be written in the form

$$\pi_{ij}(t) = \mathfrak{I}_{ij}(t) + \sum_{k=1}^{N} p_{kj}\int_{u=0}^{t} h_{jk}(t-u)\mathfrak{I}_{kj}(u)du$$

$$(i = 1, 2, \ldots N; j = 1, 2, \ldots N)$$

These equations can be expressed in matrix notation as

$$\boldsymbol{\pi}(t) = \mathfrak{I}(t) + \int_{u=0}^{t} \{\mathbf{p} \otimes \mathfrak{I}(u)\}\mathbf{h}(t-u)du$$

where $\boldsymbol{\pi}$, $\mathfrak{I}$, $\mathbf{p}$ and $\mathbf{h}$ are $N \times N$ matrices, and the symbol $\otimes$ signifies that element (i, j) in the matrix within $\{.\}$ is p_{ij} multiplied by $\mathfrak{I}_{ij}(u)$.

Cox and Miller (1965) show how the method of Laplace transforms may be used to obtain $\pi^*(s)$, the Laplace transform of $\pi(t)$. If each of the $N \times N$ equations for $\pi_{ij}(t)$ is multiplied by $\exp(-st)$ and integrated with respect to t from zero to infinity, the following equation is obtained:

$$\pi^*(s) = \mathbf{G}^*(s) + \mathbf{h}^*(s)\mathbf{G}^*(s)$$

where in general

$$\mathbf{G}^*(s) = \mathbf{p} \otimes \mathfrak{I}^*(s)$$

or, for the special case $N = 2$,

$$\mathbf{G}^*(s) = \begin{bmatrix} p_{11}\mathfrak{I}^*_{11}(s) & p_{12}\mathfrak{I}^*_{12}(s) \\ p_{21}\mathfrak{I}^*_{21}(s) & p_{22}\mathfrak{I}^*_{22}(s) \end{bmatrix}.$$

Denoting by $\mathbf{g}^*(s)$ the matrix of Laplace transforms

$$\mathbf{g}^*(s) = \mathbf{p}\otimes\mathbf{f}^*(s)$$
$$= \begin{bmatrix} p_{11}f^*_{11}(s) & p_{12}f^*_{12}(s) \\ p_{21}f^*_{21}(s) & p_{22}f^*_{22}(s) \end{bmatrix}$$

when $N = 2$, Cox and Miller (1965) show that $\pi^*(s)$ is given by

$$\pi^*(s) = s^{-1}(\mathbf{I} - \mathbf{g}^*(s))^{-1}(\mathbf{p} - \mathbf{g}^*(s)),$$

whence $\pi(t)$ may be obtained, either directly when the Laplace transforms are of easily recognised form, or by numerical integration otherwise.

Although, in the application by Rodriguez-Iturbé and Valdes (1979) the matrix $\mathbf{p}$ has many zero elements, some severe simplifications were necessary before they could proceed to make use of this general form for relating the instantaneous unit hydrograph of a basin to geomorphological characteristics. They assumed (1) that the renewal density functions $h_{ij}(t)$ (which they termed the "holding time distributions", since they represent the distributions of times of transition to sub-basin j, conditional on the particle of rainfall excess being in state i at time $t = 0$) did not depend on j, so that $h_{ij}(t) = h_i(t)$; (2) that each $h_i(t)$ was an exponential pdf: $h_i(t) = \lambda_i \exp(-\lambda_i t)$. Then, for a third-order basin ($N = \Omega + 1 = 4$), they

(1) expressed the instantaneous unit hydrograph in terms of the $\pi_i(t)$. If $A_i (i = 1 \ldots 4)$ are the areas sub-basins of order i within the total basin area A ($A_1 + A_2 + A_3 + A_4 = A$) then the instantaneous unit hydrograph is

$$(A_1/A)\pi_1(t) + (A_2/A)\pi_2(t) + (A_3/A)\pi_3(t) + (A_4/A)\pi_4(t);$$

(2) related the non-zero transition probabilities p_{12} and p_{13} in terms of the Horton numbers R_A, R_B and R_L,

(3) related the parameters λ_1, λ_2 and λ_3 in terms of an internal scale parameter L_Ω, and a mean velocity of streamflow v.

In summary, the basis underlying this remarkable approach is (1) a Markov chain of transition probabilities $[p_{ij}]$ giving the probabilities that a particle of rainfall excess moves from sub-basin i to sub-basin j; (2) the pdfs $f_{ij}(t)$, of times spent in making the transition from sub-basin i to sub-basin j; (3) the conditional pdfs that the particle of rainfall excess is in sub-basin j at time t, given that it was in sub-basin i at time $t = 0$.

6.6 MODELLING SEQUENCES OF HOURLY RAINFALL

The Stern–Coe models of Chapter Two were models of daily rainfall, and cannot be used to infer characteristics of rainfall at shorter time periods.

However, in practice some problems can only be tackled if rainfall data exist for shorter time intervals: for example urban drainage systems must be designed to deal with heavy rain falling on largely impermeable surfaces during periods of an hour or less; comparisons of forage conservation procedures by modelling require hourly values of meteorological variables (Glasby, 1995); rainfall–runoff modelling of small drainage basins commonly require rainfall data for time intervals less than a day. In addition to such practical problems, rainfall as a hydrological process is of theoretical interest, since its amount and distribution in time help shape the landscape and constitute the driving variable in all hydrological study.

Theoretical study of rainfall as a stochastic process has well over 30 years of history, but has developed rapidly in the last decade, perhaps in consequence of the increased availability of radar images showing storm position at closely spaced points in time. Waymire and Gupta (1981a, b ,c) published a series of three papers which stimulated much subsequent work, and more recently effort has been concentrated on variants of two particular models mentioned in Chapter Two: namely, the **Bartlett–Lewis model**, and the **Neyman–Scott model**. Both describe the clustering of rain "cells" using a Poisson process to represent the occurrence of clusters; in both, the number of cells per cluster is a random variable; but the two models differ in the way in which cells are distributed within each cluster.

The Bartlett–Lewis model, in its general form, has the following construction:

1. The times $t_1, t_2, t_3 \ldots$ at which clusters originate are determined by a Poisson process.
2. The numbers $N_1, N_2, N_3, \ldots$ of events per cluster are independently and identically distributed random variables from some appropriate discrete pdf.
3. The intervals between successive events in a cluster are independently and identically distributed with pdf $f(t)$, beginning at the times $t_1, t_2, t_3 \ldots$ established in (1). Thus, the N_i events in cluster i form a finite renewal process, terminated by a mechanism operating in terms of numbers of points rather than in time. An event may or may not be sited at the times $t_1, t_2, t_3 \ldots$ at which clusters originate. In the Neymann–Scott process, on the other hand, step (3) is replaced by (3*) as follows:

(3*) The times at which the N_i events in the ith cluster occur are determined by sampling from a pdf $f(t)$. For example, the times at which the N_i events occur might be determined by sampling from a normal distribution centred on the cluster time t_i; some of the events would then occur before the cluster time, as well as after it. As another example, the N_i events might be determined by sampling from an exponential distribution defined on the interval extending from the cluster time, t_i, to infinity; all events would then occur at times greater than the cluster time, but the time order in which events were marked on the time axis need not bear any relation to the order which really occurred.

4 In specific applications of both models, the general forms have been modified in various ways. For example, in to the modelling of hourly rainfall data described by Rodriguez-Iturbé et al. (1987, 1988) and Khaliq and Cunnane (1996),

1. The storm arrival times (which initiate clusters of raincell events) were determined from a Poisson process with rate parameter λ.
2. Each storm origin was followed by a Poisson process of raincell origins with rate parameter β.
3. The process of new cell origins terminated after a time exponentially distributed with parameter γ.
4. The duration of each cell was exponentially distributed with parameter η.
5. Each cell depth X was a random constant (for the duration of the cell's "life") having an exponential distribution with mean μ_X.

The number C of raincells per storm thus had a geometric distribution with mean $\mu_c = 1 + \kappa/\phi$, where $\kappa = \beta/\eta$ and $\phi = \gamma/\eta$ are dimensionless parameters introduced by Rodriguez-Iturbé et al. (1987). The model is then determined by a set of five parameters: namely $\{\lambda, \beta, \gamma, \eta, \mu_X\}$. When this form of the Bartlett–Lewis model was tested using hourly data from Denver, Colorado, Rodriguez-Iturbé at al. (1987) found that sequences generated from the fitted model gave summary statistics of rainfall accumulated over periods from 1 to 24 hours which agreed well with corresponding statistics calculated from the record; extreme values in the generated and observed sequences also agreed well. In the five-parameter model, however, there was poor agreement between the frequencies of dry spells of duration greater than one hour, when the generated sequences were compared with the recorded sequence. Rodriguez-Iturbé et al. (1988) therefore introduced a further parameter, α, by allowing the duration η of each cell depth to have a gamma distribution with index α and scale parameter v; the mean and variance of the gamma distribution were therefore α/v and α/v^2 respectively, and η varied randomly from storm to storm. The six parameters were then $\{\lambda, \kappa, \phi, v, \alpha, \mu_X\}$.

At any point in time t, the rainfall intensity is $Y(t)$, the sum of the intensities associated with all cells that are "alive" at time t. However, since rainfall is collected at discrete time intervals, here assumed to be hourly, the aggregated process Y_i is of greater interest, where Y_i is the accumulated rainfall over the ith hour; and since, more generally, properties of accumulated rainfall over periods of h hours need to be studied, Y_i^h is used to denote the rainfall total in the ith interval of duration h hours. Rodriguez-Iturbé et al. (1987) give the means, variances and covariances of the aggregated sequence Y_i^h, for any period h of aggregation, as follows:

$$\mathrm{E}[Y_i^h] = \lambda h v \mu_X \mu_c/(\alpha - 1) \tag{6.18}$$

$$\begin{aligned}\mathrm{var}[Y_i^h] = {} & 2A_1[(\alpha - 3)hv^{2-\alpha} - v^{3-\alpha} + (v + h)^{3-\alpha}] \\ & - 2A_2[\phi(\alpha - 3)hv^{2-\alpha} - v^{3-\alpha} + (v + \phi h)^{3-\alpha}]\end{aligned} \tag{6.19}$$

and for $k \geq 1$

$$\begin{aligned}\mathrm{cov}[Y_i^h, Y_{i+k}^h] = {} & A_1\{[v + (k + 1)h]^{3-\alpha} - 2(v + kh)^{3-\alpha} \\ & + [v + (k - 1)h]^{3-\alpha}\} - A_2\{[v + (k + 1)\phi h]^{3-\alpha} \\ & - 2(v + k\phi h)^{3-\alpha} + [v + (k - 1)\phi h]^{3-\alpha}\},\end{aligned} \tag{6.20}$$

where

$$A_1 = \lambda\mu_c v^\alpha[2\mu_X^2 + \kappa\phi\mu_X^2/(\phi^2 - 1)]/\{(\alpha - 1)(\alpha - 2)(\alpha - 3)\}$$
$$A_2 = \lambda\mu_c\kappa\mu_X^2 v^\alpha/\{\phi^2(\phi^2 - 1)(\alpha - 1)(\alpha - 2)(\alpha - 3)\}.$$

The probability that a period of length h is dry is then (Rodriguez-Iturbé et al., 1988)

$$\begin{aligned}P[Y_i^h = 0] = {} & \exp\{-\lambda h - \lambda\mu_T + \lambda G_P^*(0,0) \\ & \times [\{\phi + \kappa v^{\alpha-1}/[v + (\kappa + \phi)h]^{\alpha-1}\}/(\phi + \kappa)]\}\end{aligned} \tag{6.21}$$

where μ_T is the expected duration of a single storm, for which Rodriguez-Iturbé at al. (1987) give an exact but lengthy expression; an approximation is

$$\begin{aligned}\mu_T \approx {} & v/[\phi(\alpha - 1)].[1 + \phi(\kappa + \phi) - \phi(\kappa + \phi)(\kappa + 4\phi)/4 \\ & + \phi(\kappa + \phi)(4\kappa^2 + 27\kappa\phi + 72\phi^2)/72].\end{aligned}$$

An approximation to $G_P^*(0, 0)$ is

$$G_P^*(0, 0) \approx v/[\phi(\alpha - 1)](1 - \kappa - \phi + 3\kappa\phi/2 + \phi^2 + \tfrac{1}{2}\kappa^2).$$

A problem of considerable practical difficulty is the estimation of the six parameters $\{\lambda, \kappa, \phi, v, \alpha, \mu_X\}$. Complexity of the model structure does not permit (as yet) parameters to be estimated by maximum likelihood; consequently, estimation is by the method of moments. To estimate the six parameters, six characteristics of the data record (which might be means, variances or covariances for varying values of h: $h = 1, 6, 24, \ldots$ for example; or proportions of intervals for $h = 1, 6, 24, \ldots$ that are dry) are equated to the expressions for means, variances and covariances given in (6.18), (6.19) and (6.20), or to the probability that a period of length h is dry (the expression in (6.21)). Thus, six equations are obtained to be solved for the unknown parameters $\{\lambda, \kappa, \phi, v, \alpha, \mu_X\}$. In fact, it is possible to use more than six equations; if more are used, estimates of the six parameters are obtained which are solutions of the equations in a least-squares sense: given p equations $g_i = 0$ for $i = 1, 2, \ldots p$, the values of $\lambda, \kappa, \phi, v, \alpha, \mu_X$ are calculated which minimise the expression

$$\sum_{i=1}^{p} g_i(\lambda, \kappa, \phi, v, \alpha, \mu_X)^2. \tag{6.22}$$

Stability and sensitivity of estimates of the six parameters can depend upon which of the many possible equations $g_i = 0$ are selected. Khaliq and Cunnane (1996) studied this problem using hourly rainfall records from Valentia and Shannon airports in Ireland, the former being relatively wet and the latter relatively dry. They concluded that estimates were most stable and least sensitive when the six parameters were obtained using $p = 16$ in (6.22); the 16 equations used were (1) four means ((6.18) used for $h = 1, 6, 12$ and 24 hours); (2) four variances ((6.19) used for $h = 1, 6, 12$ and 24 hours); (3) four lag-one autocovariances (derived from (6.20) and (6.19), for $k = 1$ and $h = 1, 6,$ 12 and 24 hours); and (4) four proportions of dry periods ((6.21) for $h = 1, 6, 12$ and 24 hours).

Modifications allowing for η to vary from storm to storm, according to a gamma distribution with parameters α and v, have also been studied for the Neyman–Scott model by Entekhabi et al. (1989); Cowpertwait (1991) derived some additional properties for this model. In contrast to the Bartlett–Lewis model, whose use is restricted to one dimension, the Neyman–Scott model extends easily to two and three dimensions, and can therefore be used, for example, to model the development of storm cells in two-dimensional space.

6.6.1 Disaggregation of daily rainfall

Whilst records of daily rainfall are relatively common in many parts of the world, records of hourly rainfall are far less common. The question therefore arises: is it possible to use records of hourly rainfall, supposing that a suitable model can be found to describe them, to "disaggregate" totals in the more plentiful records of daily rainfall? Glasbey et al. (1995) have used the Bartlett–Lewis model of hourly rainfall as modified by Rodriguez-Iturbé et al. (1987, 1988) to study this problem; their purpose was to use the disaggregated records to model the drying of grass swathes in field pastures.

The first step was to fit the Bartlett–Lewis model to 10 years of hourly rainfall (1968–77) at Edinburgh airport (Turnhouse), Scotland; their fitting procedure used six statistics corresponding to the mean, variance, lag-one serial correlation coefficient and the proportion of wet periods, for $h = 1$ hour, and the variance and proportion of wet periods for $h = 24$ hours. The estimates of the six parameters α, κ, λ, μ_X, v and ϕ were 2.56, 0.455, 0.0268, 1.10, 0.510 and 0.0751 respectively.

To disaggregate daily rainfall data, it is necessary to simulate series of hourly data using the fitted model which are constrained to have the appropriate daily totals. This is not at all straightforward as there is no known expression for the distribution of hourly totals, conditional on the daily totals being given. Therefore, the necessary alternative used by Glasbey et al. (1995) was to generate a sequence of hourly rainfalls unconditionally, using the fitted model, until a rainfall sequence was generated having the correct daily totals. The model was used to generate 1000 years of hourly "data"; patterns of rain were then matched between observed and simulated daily totals, and hourly values for the best match were used for disaggregation.

A different approach to the problem of disaggregating daily rainfall was given by Econopouly et al. (1990). These authors did not seek to disaggregate daily totals into hourly sub-totals, but into sequences of within-day shower totals. They assumed a joint distribution for the number of showers N, and for the daily rainfall total Z' (since the lower limit of rainfall measurement was 0.254 mm, Z' was rainfall depth minus this threshold). They therefore wrote the joint distribution $H(n, z)$ of the two variables N, Z' as

$$H(n, z) = G(n \mid z)F(z),$$

where $F(z)$ is the marginal distribution of the variable Z'. The forms for $F(z)$ and $G(n \mid z)$ had then to be selected; for $F(z)$, the alternatives tested were the exponential, gamma, mixed exponential and Weibull distributions, and it was found that in general the mixed exponential pdf

$$f(z) = \alpha/\beta \exp(-z/\beta) + (1-\alpha)/\theta \exp(-z/\theta)$$

performed most satisfactorily. For the conditional distribution $G(n \mid z)$ of number of within-day showers given daily rainfall depth, a modified form of the negative binomial distribution (see Chapter Two) was used in which the two parameters varied with rainfall depth. Further distributions were assumed for the ratios of within-day shower rainfalls to daily totals, and for the joint distribution of shower duration and depth. The method was based on earlier work by Hershenhorn (1984) and Hershenhorn and Woolhiser (1987). An important characteristic of their method is that transferability of disaggregation parameters over fairly large distance within climatologically homogeneous regions was shown to be feasible, at least in the Midwestern United States.

6.6.2 Fitting the modified Bartlett–Lewis model: an example

This example, kindly supplied by Back (1997), uses (1) hourly rainfall for the period October 1980 to December 1996, obtained by a recording gauge; and (2) records from a daily raingauge near by, sited at the FEPAGRO experimental station (FEPAGRO: Funclacáo Esta dual de Pesquisas em Agropecuária) at Urussanga in the State of Santa Catarina, southern Brazil. The purpose of the calculation to be described was to obtain a method for disaggregating daily rainfall into hourly components for use in a detailed water-balance study of crops receiving irrigation to supplement rainfall.

The first step was to separate the entire record of hourly rainfall into months, so that the Januarys, Februarys, could be treated separately and parameters estimated for each month of the year. Taking each month separately (the Januarys, for example) further series were derived by accumulating the hourly rainfalls into totals for 6, 12 and 24 hours. Thus for each month,

Table 6.2 Estimates of the six parameters λ, ν, κ, κ, μ_X, α, ϕ of the modified Bartlett–Lewis model: hourly rainfall 1980–96, Urussanga, SC, Brazil

Month	λ	ν	κ	μ_X	α	ϕ
Jan.	0.0244	1.096	0.084	11.40	5.57	0.0263
Feb.	0.0172	0.671	0.043	11.29	2.87	0.0120
Mar.	0.0224	1.670	0.091	7.87	5.88	0.0443
Apr.	0.0172	0.474	0.161	7.82	3.27	0.0353
May	0.0071	0.557	0.164	4.10	2.01	0.0179
June	0.0118	0.833	0.954	1.92	2.38	0.1059
July	0.0116	2.760	1.180	1.48	4.02	0.1311
Aug.	0.0104	1.279	0.381	3.19	4.52	0.0366
Sept.	0.0165	0.818	0.376	2.36	2.33	0.0605
Oct.	0.0214	0.331	0.466	4.58	2.83	0.0486
Nov.	0.0176	0.645	0.130	7.46	4.08	0.0242
Dec.	0.0176	0.382	0.100	10.64	2.88	0.0161

four series were obtained with rainfall accumulated over h hours, where $h = 1, 6, 12, 24$.

For each value of h, means, variance, covariances and correlations were calculated. Thus if the subscript $k (k = 1, 2, \ldots 12)$ indicates month of the year, the following quantities were estimated:

1. $\mu_k(h)$, where $\mu_k(h)$ is the mean rainfall when it is accumulated over periods of h hours in month k;
2. $\gamma_k(h)$, where $\gamma_k(h)$ is the variance of rainfall when it is accumulated over periods of h hours in month k;
3. $\gamma_k(h, 1)$, where $\gamma_k(h, 1)$ is the "lag-1 covariance" (the covariance calculated between a sequence of rainfall totals and the same sequence shifted one time-step to the right) computed using the sequence of rainfalls accumulated over h-hour periods in month k.

These estimates, which are statistics calculated from data, were then equated to their expected values given in (6.18), (6.19), (6.20) and (6.21) above. Their expected values were functions of the six parameters λ, ν, κ, μ_X, α and ϕ modified Bartlett–Lewis model; to estimate these six parameters, a sum of squares criterion was minimised, giving the estimates of the six parameters given in Table 6.2.

Having obtained the estimates shown in Table 6.2, the next step was to use them to see how well the disaggregation procedure (daily rainfall to hourly) performed. For each month of the year (January, for example) 500 months of hourly rainfall were generated, using the appropriate estimates of the six parameters of the modified Bartlett–Lewis model taken from Table 6.2. The procedure then followed that of Glasbey et al. (1995) described above; days in the generated sequence were identified which had a daily total nearest to the recorded daily rainfall. Having identified this day in the generated sequence,

Table 6.3 Comparison of statistics calculated from (a) hourly rainfall, (b) disaggregated daily rainfall, 1980–96 at Urussanga, SC, Brazil. $\mu_k, \gamma_k(1), \gamma_k(1,1), \gamma_k(6), \gamma_k(6,1), \gamma_k(12), \gamma_k(12,1)$, are as defined in the text; $pd(1)$, $pd(6)$ and $pd(12)$ are proportions of periods of length 1, 6 and 12 hours that are dry

	μ_k		$\gamma_k(1)$		$\gamma_k(1,1)$		$\gamma_k(6)$		$\gamma_k(6,1)$	
	(a)	(b)	(a)	(b)	(a)	(b)	(a)	(b)	(a)	(b)
Jan.	0.28	0.28	2.79	2.76	0.80	0.66	30.25	28.91	6.76	5.42
Feb.	0.32	0.32	3.79	3.58	1.48	1.37	51.18	44.80	12.23	18.46
Mar.	0.18	0.18	1.49	1.72	0.51	0.45	16.98	15.50	3.16	3.20
Apr.	0.16	0.16	1.16	1.19	0.43	0.40	15.19	13.99	4.25	3.63
May	0.16	0.16	0.97	0.97	0.59	0.55	19.38	17.35	9.05	7.55
June	0.14	0.14	0.57	0.61	0.36	0.40	12.20	12.22	6.04	5.85
July	0.16	0.16	0.59	0.63	0.38	0.44	12.24	13.30	6.47	4.91
Aug.	0.14	0.14	0.60	0.70	0.30	0.36	9.94	10.89	3.79	3.62
Sept.	0.17	0.17	0.68	0.72	0.42	0.46	13.89	13.10	6.47	5.35
Oct.	0.19	0.19	1.01	0.94	0.49	0.42	16.44	14.10	5.93	5.34
Nov.	0.18	0.18	1.17	1.24	0.39	0.39	13.84	13.60	3.84	3.87
Dec.	0.27	0.27	2.62	2.47	0.95	0.83	33.78	29.93	8.98	9.05

	$\gamma_k(12)$		$\gamma_k(12,1)$		$pd(1)$		$pd(6)$		$pd(12)$	
	(a)	(b)	(a)	(b)	(a)	(b)	(a)	(b)	(a)	(b)
Jan.	71.7	69.4	11.7	12.4	0.87	0.91	0.75	0.75	0.64	0.65
Feb.	136.4	128.6	14.1	35.0	0.86	0.91	0.72	0.72	0.60	0.60
Mar.	39.6	38.7	5.2	5.2	0.90	0.93	0.80	0.80	0.69	0.71
Apr.	36.6	35.3	8.0	9.1	0.91	0.93	0.82	0.84	0.74	0.76
May	55.7	48.2	21.4	22.1	0.90	0.93	0.84	0.84	0.78	0.78
June	34.9	37.1	15.6	13.5	0.90	0.93	0.84	0.86	0.77	0.80
July	37.6	37.9	13.4	10.7	0.89	0.91	0.83	0.85	0.77	0.78
Aug.	26.2	31.2	10.3	7.9	0.90	0.92	0.83	0.85	0.77	0.80
Sept.	39.4	37.7	13.9	11.6	0.87	0.90	0.78	0.80	0.70	0.72
Oct.	44.7	39.4	12.4	11.2	0.88	0.91	0.78	0.82	0.69	0.72
Nov.	36.2	34.5	7.6	8.0	0.88	0.92	0.78	0.80	0.69	0.72
Dec.	84.2	78.8	21.5	25.7	0.88	0.90	0.77	0.78	0.68	0.68

the rainfall in its 24-hour sequence was taken as the disaggregated sequence for the observed record. Table 6.3 compares the statistics of the observed record of hourly rainfall at Urussanga, with the same statistics calculated from the disaggregated record obtained from the other (daily) gauge near by. Agreement between the statistics is generally good, suggesting that the disaggregation procedure is a satisfactory basis for subsequent water-balance studies of irrigated crops.

6.7 THE SHEN–TODOROVIC MODEL OF BEDLOAD PARTICLE MOVEMENT

It has long been recognised that the transport of bedload material in rivers could be regarded as a random process in which each individual particle proceeds downstream in a series of jumps ("saltations") separated by periods of immobility. Einstein (1937) is quoted (Shen and Todorovic, 1971) as having developed the first stochastic model of how such particles move, based on the assumptions that (1) the velocity field was stationary in time and homogeneous in the lateral as well as the longitudinal directions; (2) the lengths of the transport periods were negligibly small relative to the lengths of the periods of immobility; (3) the probability that a particle moved was independent of its location. Einstein, and others working on similar lines concluded from a theoretical analysis that both the step lengths and the rest periods were distributed exponentially, a conclusion not borne out by subsequent experimental study (Shen and Todorovic, 1971). These authors presented a general stochastic model for the transport of sediment bed material which is in fact a time-dependent Poisson process. Here, we use a notation simpler than that of Shen and Todorovic, and begin by introducing the terminology.

6.7.1 Notation

We consider a typical sediment particle, alternating between periods of movement and immobility (noting that these characteristics suggest a treatment in terms of the alternating renewal process described earlier in this chapter, although this was not the approach of Shen and Todorovic). It is assumed that the particle always moves in the downstream direction, although there are local circumstances – particles moving in the wake of a dune – for which the assumption may not hold, although the assumption will in general be true. We consider the events "E: the particle makes exactly n steps in the time period $(0, t]$" and "G: the particle makes exactly n steps in the distance $(0, x]$". Since the probabilities of these events will depend on t and x respectively, we write $p_n(t)$ for the probability of the event E, and $r_n(x)$ for the probability of the event G.

Central to the development of Shen and Todorovic is the Poissonian assumption demonstrated as follows. Consider development in the time dimension; the argument in the space dimension is identical. Suppose we take a new time origin at t_0 instead of $t = 0$; t_0 may be a point at which an event has occurred, or any other fixed point in time. Suppose that the first event after t_0 occurs at $t_0 + Z$; we consider the probability $p_1(t^*) = P[Z > t^*]$. Then

$$p_1(t^* + \Delta t^*) \equiv P[Z > t^* + \Delta t^*]$$

must equal the probability that $Z > t^*$ and that no event occurs in the interval $(t_0 + t^*, t_0 + t^* + \Delta t^*]$; but, by the definition of conditional probability,

$$P[(Z > t^*) \cap (\text{no event occurs in the interval } (t_0 + t^*, t_0 + t^* + \Delta t^*])] \\ = P[(Z > t^*)].P[(\text{no event occurs in the interval } \\ (t_0 + t^*, t_0 + t^* + \Delta t^*]) \mid Z > t^*]. \quad (6.23)$$

Equation (6.23) holds for any process of events; however, if the final probability in (6.23) also satisfies

$$P[(\text{no event occurs in the interval } (t_0 + t^*, t_0 + t^* + \Delta t^*]) \mid Z > t^*] \\ = P[(\text{no event occurs in the interval } (t_0 + t^*, t_0 + t^* + \Delta t^*])], \quad (6.24)$$

then we have a restatement of the Markov property that what happens in the interval $(t_0 + t^*, t_0 + t^* + \Delta t^*]$ is independent of what happened at or before $t_0 + t^*$. This Markovian property is central to the model of Shen and Todorovic, although they present it in a slightly different form: namely, that the probability of a single step in $(t, t + \Delta t]$ conditional on n steps having occurred in $(0, t]$, is equal to the unconditional probability of a single step in $(t, t + \Delta t]$.

Following notation in earlier chapters, we represent by $\lambda(t)$ the rate at which the particle's steps occur at time t, so that the probability of a step occurring in a short interval of time Δt is $\lambda(t).\Delta t + o(\Delta t)$. The probability that no step occurs in Δt is

$$(1 - \lambda(t).\Delta t) + o(\Delta t),$$

and the probability that two or more steps occur in Δt is $o(\Delta t)$.

The argument has been in terms of the process developing in time t; the same argument holds for the development of events in the space indexed by x. Assuming this Markov property, the Shen–Todorovic model can be developed as shown in the next section.

6.7.2 Equations for particle movement in time and space

Recall that $p_n(t)$ is the probability that a particle makes exactly n steps in the interval $(0, t]$; consider $p_n(t + \Delta t)$, where Δt is sufficiently small that no more than one step is made during its duration. This is the probability that the particle makes exactly n steps in $(0, t + \Delta t]$, an event which can occur in either of two mutually exclusive ways: either (1) the particle had made n steps in $(0, t]$, and no further step was taken in $(t, t + \Delta t]$, or (2) it had taken $n - 1$ steps in $(0, t]$, and an additional step was taken in $(t, t + \Delta t]$.

Since, as shown in the preceding section, the events "n steps in $(0, t]$" and "no further step was taken in $(t, t + \Delta t]$" are independent, their probabilities multiply, giving

$$p_n(t + \Delta t) = \{1 - \lambda(t)\Delta t\}.p_n(t) + \lambda(t)\Delta t.p_{n-1}(t) \quad (6.25)$$

ignoring terms in Δt higher than the first. A rearrangement gives

$$\{p_n(t + \Delta t) - p_n(t)\}/\Delta t = \lambda(t)\{p_{n-1}(t) - p_n(t)\}$$

and as Δt tends to zero, we have

$$\mathrm{d}p_n(t)/\mathrm{d}t = \lambda(t)\{p_{n-1}(t) - p_n(t)\} \tag{6.26}$$

for $n = 1, 2, 3, \ldots$. For the case $n = 0$, (6.26) becomes

$$p_0(t + \Delta t) = \{1 - \lambda(t)\Delta t\}.p_0(t) \tag{6.27}$$

since $p_{-1}(t) = 0$ (a negative number of steps has probability zero). Hence (6.27) gives

$$\mathrm{d}p_0(t)/\mathrm{d}t = -\lambda(t)p_0(t). \tag{6.28}$$

The equations (6.26) and (6.28) constitute a set of non-linear differential equations which can be solved in a number of ways. A recursive solution is one possibility (solve (6.28) explicitly for $p_0(t)$; substitute in (6.26) for $n = 1$; repeat). Another alternative is to use probability generating functions: multiply (6.26) by z^n, add over n to give

$$\mathrm{d}G(z, t)/\mathrm{d}t = \lambda(t)(z - 1)G(z, t),$$

where $G(z, t) = \sum p_n(t)z^n$. Integration of this equation gives

$$G(z, t) = \exp(-\Lambda(t)).\exp(-\Lambda(t)z), \tag{6.29}$$

where

$$\Lambda(t) = \int_{u=0}^{t} \lambda(u)\mathrm{d}u. \tag{6.30}$$

The function $G(z, t)$ is already familiar as the probability generating function of a time-dependent Poisson process; the coefficient in the expansion in powers of z gives

$$p_n(t) = \exp(-\Lambda(t)).\{\Lambda(t)\}^n/n! \tag{6.31}$$

as the probability that the particle makes n steps in the time interval $(0, t]$.

An exactly similar argument gives, for the probability of n steps in the distance interval $(0, x]$,

$$r_n(x) = \exp(-\varpi(x)).\varpi(x)^n/n!, \tag{6.32}$$

where

$$\varpi(x) = \int_{u=0}^{x} \lambda^*(u)\mathrm{d}u$$

and $\lambda^*(x)$ is the space rate of step occurrence, corresponding to $\lambda(t)$.

Now let X_t denote the position of the particle at time t, after it has taken n_t steps. The distribution function of X_t is

$$\begin{aligned} F_t(x) &= P[X_t \le x] \\ &= \sum_{j=0}^{\infty} P[(X_t \le x) \cap (n_t = j)] \\ &= \sum_{j=0}^{\infty} P[(X_t \le x) \mid (n_t = j)].P[n_t = j] \end{aligned}$$

by the theorem of total probability; hence

$$F_t(x) = \exp\{-\varpi(x) - \Lambda(t)\}.\sum_{j=0}^{\infty}\sum_{k=j}^{\infty}[\varpi(x)]^k.[\Lambda(t)]^j/[k!j!]. \qquad (6.33)$$

Differentiation with respect to x gives the pdf $f_t(x)$ of the particle's position.

In practical terms, the difficulty is to identify the functions $\varpi(x)$ and $\Lambda(t)$, and much effort has been directed to establishing these functions through study of particle behaviour in experimental flumes; for example Vukmirovic and Wilson, Jr. (1976, 1981) conducted an extensive series of flume studies with radioactively labelled sand grains, concluding that the rates $\lambda(t)$ and $\lambda^*(x)$ depended not only on t and x respectively, but also on n_t according to the following forms:

$$\begin{aligned} \lambda(t, n_t) &= \{\alpha + \beta\exp(-\beta t)\}.\{1 + \theta n_t/(1 + n_t)\} \\ \lambda^*(x, n_t) &= a(1 - \exp(-bx)\}/\{1 + n_t/\rho\}, \end{aligned}$$

where $\alpha, \beta, \theta, a, b$ are positive constants and ρ is particle density.

6.7 METHODS FOR TESTING VALIDITY OF THE POISSON ASSUMPTION

Tests that data conform to a Poisson process are based on the distributional properties either of the numbers of event occurrences, as in Chapter Two, or of the intervals between events. Cox and Lewis (1966) give a very full account of test procedures based not only on distributions of numbers of events and of intervals between them, but also including periodogram analysis of the numbers of events and their inter-event distances. We note here that a non-homogeneous Poisson process with rate $\lambda(t)$ can be transformed (Cox and Miller, 1965) to a simple Poisson process by a transformation of the time scale to a new time variable τ. The transformation $\tau(t)$ must be a non-decreasing function of t, chosen such that

$$\lambda(t)\mathrm{d}t/\mathrm{d}\tau = 1$$

so that

$$\tau(t) = \int_{u=0}^{t} \lambda(u)\mathrm{d}u = \Lambda(t).$$

With this transformation, the non-homogeneous process with

$$P[N = n] = \exp(-\Lambda(t))\{\Lambda(t)\}^n/n!$$

becomes

$$P[N = n] = \exp(-\tau)\tau^n/n!$$

namely, a simple Poisson process with unit rate. Time intervals $\Lambda(t_{i+1}) - \Lambda(t_i)$ between events occurring at times t_i, t_{i+1} will then follow an exponential distribution

$$f(\tau) = \exp(-\tau).$$

To illustrate some approaches, we take Hurley's (1992) data on bedload events in two streams in the uplands of northern England, Great Eggleshope Beck (period 8 December 1978–29 August 1984) and Carl Beck (5 October 1978–20 March 1984). The data are given below in Table 6.4. It will be recalled that the data were shown to come from non-homogeneous Poisson and gamma processes, and Hurley reported tests of goodness of fit for the gamma-process data by probability plotting, confirming that the gamma process fitted adequately apart from one large interval at Great Eggleshope Beck when most of the precipitation fell as snow.

The worked example presented below is less sophisticated than Hurley's, giving a more exploratory study of the data. It takes a simple Poisson process as a reasonable point of departure and then considers some of the departures from the simple Poisson suggested by the data.

First, the inter-event times are calculated by simple differencing; the mean time interval at Great Eggleshope Beck is 26.06 days, with an estimated rate (assumed constant for the present) of $1/26.06 = 0.03837$ events per day. The corresponding mean for Carl Beck is 24.68 days, with estimated rate 0.04052 events per day. If the inter-event times are exponential with constant λ, the mean and variance at each site would be $1/\lambda$ and $1/\lambda^2$, so that the standard deviations of inter-event times should not be too different from the means of inter-event times; these standard deviations are 30.37 for Great Eggleshope Beck and 29.14 for Carl Beck: rather larger than the means, so we should not be surprised to find evidence of departure from the hypothesis of a simple Poisson process. In fact the coefficients of variation at the two sites are 1.165 and 1.181.

Table 6.4 Data from M. Hurley (1992): *J. Hydrol.* **138**, 540–1. Data from Great Eggleshope Beck: days of bedload events after 8 December 1978

0	17	81	85	91	107
114	124	149	163	170	292
300	309	321	330	338	343
352	359	360	361	369	374
384	392	418	428	441	459
474	480	545	554	564	570
600	608	630	680	710	713
717	725	734	737	746	767
774	786	820	837	872	1023
1028	1076	1081	1121	1174	1241
1302	1398	1409	1435	1444	1461
1472	1489	1546	1597	1636	1773
1827	1842	1861	1883	1935	2007

Data from Carl Beck: days of bedload events after 5 October 1978

0	8	41	49	64	80
93	119	144	151	157	166
174	187	215	225	235	356
375	391	394	402	407	416
423	424	425	429	433	436
447	456	481	491	505	523
537	541	610	618	624	629
634	664	672	732	743	771
777	781	798	820	831	838
851	884	887	899	900	936
1087	1092	1099	1149	1185	1245
1305	1360	1462	1508	1525	1553
1617	1661	1698	1829	1837	1891
1915	1932	1953	1999		

If the process is Poisson, the differences should follow a distribution function

$$F(t) = P\{T \leq t] = 1 - \exp(-\lambda t).$$

Expressed in slightly different terms, the **survival function** $S(t)$, defined by

$$S(t) = 1 - F(t)$$

is $S(t) = \exp(-\lambda t)$. Hence $\log_e S(t) = -\lambda t$; if therefore we put the N intervals (where $N = 77$ for Great Eggleshope Beck and $N = 81$ for Carl Beck) in ascending order to get $t_{(1)}, t_{(2)}, t_{(3)}, \ldots t_{(i)}, \ldots t_{(N)}$ and plot the proportion exceeding i against $t_{(i)}$, the plot should be a straight line passing through the origin with slope $-\lambda$. Figures 6.1 and 6.2 show these plots for Great Eggleshope Beck and Carl Beck respectively; the figures are very similar for the two sites, with the four largest intervals at each site departing appreciably from the remaining points. Further evidence of departure appears when an exponential

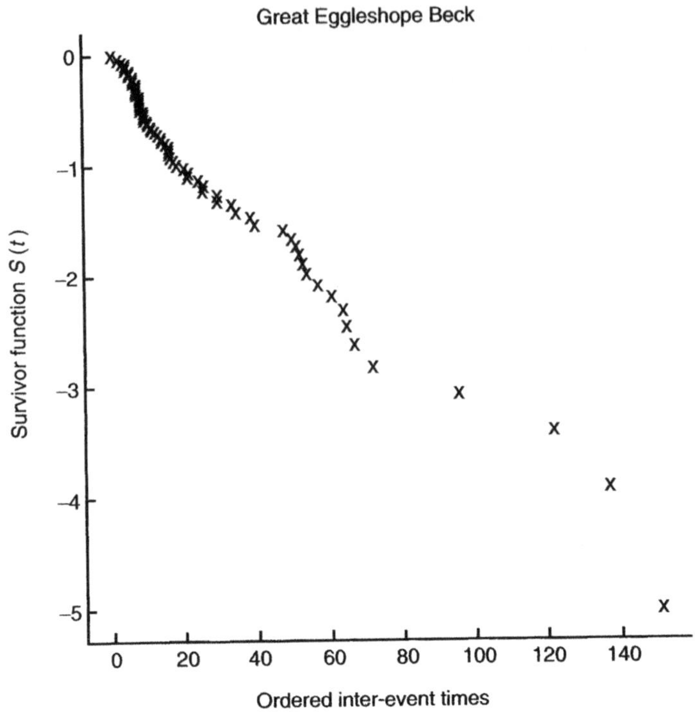

Figure 6.1 Plot of survivor function $S(t)$, Great Eggleshope Beck: a straight-line fit corresponds to a Poisson process

distribution is fitted to the intervals; if they are divided into eight classes for the purpose of fitting the distributions, deviance statistics are obtained as 20.61 on 7 df for Great Eggleshope Beck and 14.39 on 7 df for Carl Beck (the tabulated values of χ^2 for 7 df at the 5% and 1% levels of significance are 14.067 and 18.475) suggesting greater evidence of departure from the simple Poisson at Great Eggleshope Beck than at Carl Beck.

We now look at possible models for the Carl Beck data, based on the assumption of an exponential distribution with time-varying mean; we include the possibility of a time trend over the period of record, together with a seasonal periodicity in the mean. Following Hurley's approach, we consider a GLM for which (1) the expected value of each time interval is an exponential distribution with time-varying mean $\mu(t) = 1/\lambda(t)$; (2) the linear predictor is

$$\eta(t) = \beta_0 + \beta_1 \cos(2\pi t/365) + \beta_1 \sin(2\pi t/365);$$

(3) the link function is the log function

$$\eta(t) = \log\{\mu(t)\} = -\log\{\lambda(t)\},$$

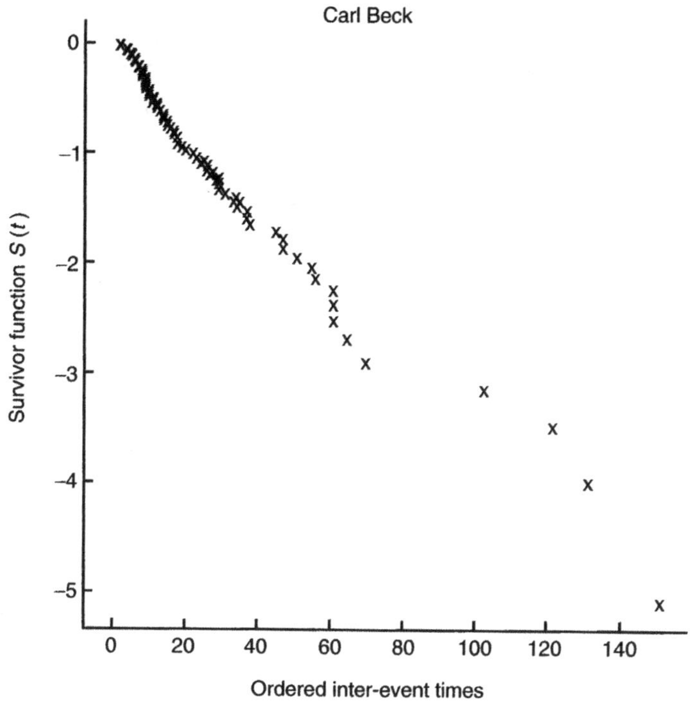

Figure 6.2 Plot of survivor function $S(t)$, Carl Beck: a straight-line fit corresponds to a Poisson process

corresponding to the model used by Hurley. The total deviance of 88.86 with 80 df is reduced, when the model is fitted, to 71.79 on 78 df, the ratio of mean deviances being (17.07/2) divided by (71.79/78) = 8.54; the tabulated 0.1% F-statistic with 2 and 78 df is about 7.5, giving strong evidence of a significant reduction in deviance and hence that the model provides a good representation of the data. However, inspection of the results shows that the final two data points influence the fit (not unexpected from Figure 6.2). The estimates of the three parameters $\beta_0, \beta_1, \beta_2$ in the linear predictor are 6.11 ± 1.47, -3.16 ± 1.23, and -1.33 ± 1.05, so that the cosine term appears the more important.

In the above analysis, the inter-event times were assumed to be exponentially distributed, the fitting procedure using the particular form of the gamma distribution in which the parameter κ was constrained to have value 1. If this restriction is removed, allowing the parameter κ to take the value suggested by the data, we find that the estimate of κ is 0.92, sufficiently near to 1 not to cause doubt about the exponential assumption. The estimates of $\beta_0, \beta_1, \beta_2$ are equal to those obtained with the exponential assumption, only their standard errors being slightly changed: namely 6.11 ± 1.41, -3.16 ± 1.18, and

Table 6.5 Numbers of bedload events occurring in 22 consecutive three-month periods: Great Eggleshope Beck and Carl Beck

Great Eggleshope Beck																					
2	4	6	1	13	6	5	3	8	5	1	1	3	2	1	1	6	2	2	0	3	3
Carl Beck																					
6	7	4	1	13	7	5	2	6	8	1	1	3	2	2	0	3	2	2	0	3	4

-1.33 ± 1.01. Recall from Chapter Two that the coefficients in the linear predictor can be estimated without the need simultaneously to estimate the dispersion parameter.

So far we have analysed the inter-event times. The numbers of events can also be analysed, and for this purpose the events were grouped into three-month periods January–March, April–June, July–September, and October–December. The data from both Great Eggleshope Beck and Carl Beck comprised 22 such three-month periods, for which the numbers of bedload events occurring are given in Table 6.5.

At Great Eggleshope Beck, the mean number of events per three-month period is 3.69 and variance 11.77, suggesting strongly that a simple Poisson process is very doubtful; for Carl Beck, the mean and variance of the numbers of events are 3.87 and 12.49. We can fit a GLM to these data also, assuming a Poisson distribution with log-link function, but the linear predictor is now modified to include a time trend, so that

$$\eta(t) = \beta_0 + \beta_1 t + \beta_2 \cos(2\pi t/4) + \beta_3 \sin(2\pi t/4),$$

remembering that there are four totals per year. With this model, the total deviance 47.76 with 21 df is reduced to 28.85 on 18 df, the mean deviance ratio being (18.90/3) divided by (28.85/18) = 6.30 (tabulated 1% F-statistic for 3 and 18 df is 5.09). Estimates of the linear predictor parameters $\beta_0, \beta_1, \beta_2, \beta_3$ are 1.652 ± 0.214, -0.0476 ± 0.0178, -0.359 ± 0.173 and 0.439 ± 0.169, confirming the negative time trend reported by Hurley.

6.8 MONTE CARLO TESTS

Monte Carlo methods provide a very general way of testing hypotheses when the distributional properties of test statistics are difficult to establish. In the case of Hurley's data, these difficulties do not arise because the distributions of test statistics such as the mean deviance are determined at least approximately from statistical theory; nevertheless we use Hurley's data to demonstrate the ease with which Monte Carlo tests can be constructed. The tests consist basically of the following steps: (1) a hypothesis is set up concerning the structure of data; (2) a test statistic is selected and its numerical value is calculated from the data; (3) on the hypothesis set up in (1), repeated samples

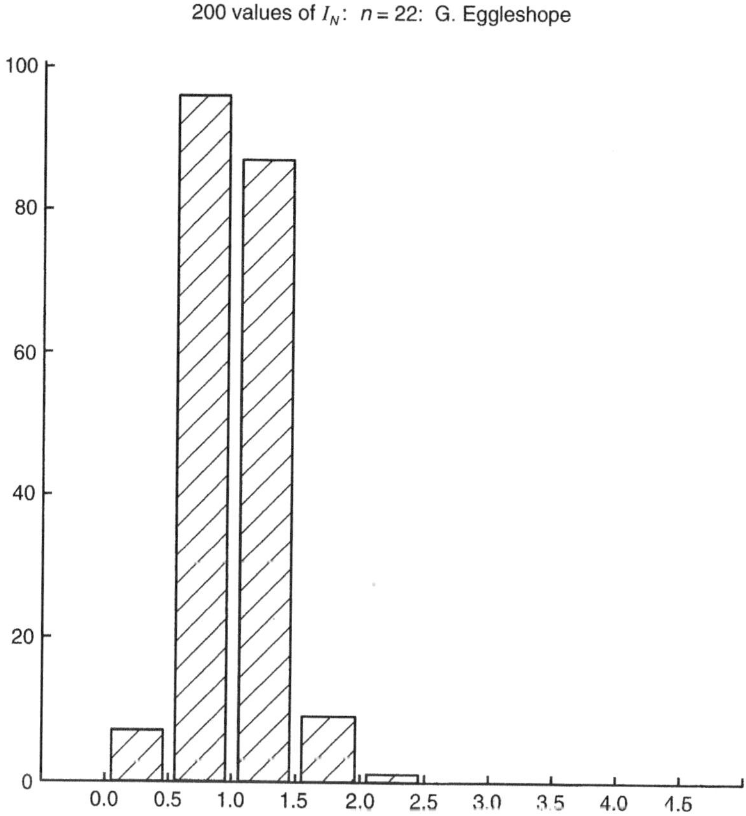

Figure 6.3 Histogram of 200 values of I_N, calculated on hypothesis that 22 counts of bedload events at Great Eggleshope Beck come from a Poisson distribution with constant λ

are generated by computer using a pseudo-random number generator; (4) the test statistic is calculated from each such generated sample, and the empirical distribution of the calculated test statistics is computed; (5) the value of the test statistic calculated from the data is compared with the empirical distribution: if its value lies far into either of the tails of the empirical distribution, then doubt is cast on the hypothesis set up in step (1). More formally, let U be the test statistic calculated from n data values and let $\{U_i: i = 2, \ldots k\}$ denote $(k-1)$ values of the test statistic generated by independently simulating data, with the same sample size $n, (k-1)$ times under the null hypothesis. Denote by U_1 the observed value of U. The k values of U (including that calculated from the data) are put into ascending order

$$U_{(1)} \leq U_{(2)} \leq U_{(3)} \leq \ldots \leq U_{(k)}$$

and the hypothesis is rejected if U_1 is one of the smaller or one of the larger

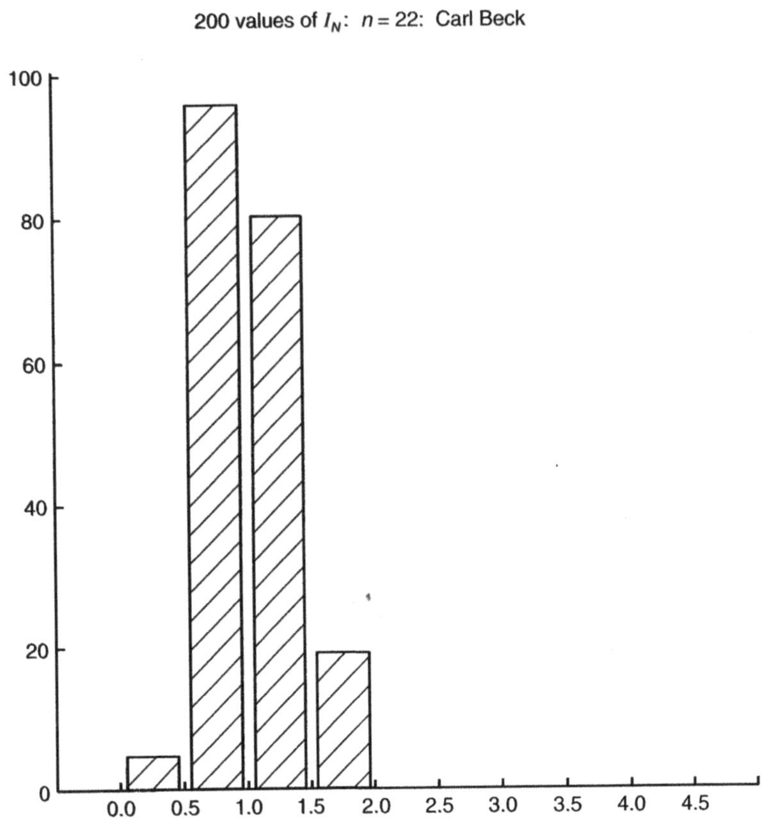

Figure 6.4 Histogram of 200 values of I_N, calculated on hypothesis that 22 counts of bedload events at Carl Beck come from a Poisson distribution with constant λ

order statistics, depending on the direction of the alternative hypothesis with regard to U.

Suppose, to illustrate, we use the 22 quarterly values given in Table 6.3 for each of the two sites Great Eggleshope Beck and Carl Beck, and wish to test the hypotheses that the 22 values at each site come from a simple Poisson process. If the data are from a Poisson process with parameter λ, the counts will have come from a a Poisson distribution with mean and variance both equal to λ, so that the ratio I_N = variance/mean should not differ very greatly from unity. However, at Great Eggleshope Beck the value of I_N is 3.190 and at Carl Beck $I_N = 3.227$, both values appreciably greater than unity. Taking I_N as the test statistic, we generate (for example) $k = 200$ samples from a Poisson distribution with mean 3.69, the mean of the 22 counts at Great Eggleshope Beck, and $k = 200$ samples from a Poisson distribution with mean 3.87, the mean of the 22 counts at Carl Beck. For each of the 200 generated samples, the test statistic I_N is calculated, and the histogram of the 200 values plotted, with the results shown in Figures 6.3 and 6.4.

Clearly the values of I_N calculated from the two data sets (3.190 and 3.227 for Great Eggleshope Beck and Carl Beck respectively) lie way into the upper tails of the two empirical distributions, giving evidence that the simple Poisson hypothesis should be rejected. The test used here was based on the test statistic I_N, but alternative test statistics are possible, such as the **Poisson index** defined as (variance − mean)/mean^2, which would have a value near to zero when the counts are Poisson. Similar tests are also possible using the inter-event times; if the event occurrences conform to a simple Poisson process, the inter-event times would have an exponential distribution with mean $1/\lambda$ and variance $1/\lambda^2$, suggesting the test statistic I_T = variance/(mean^2) which, on the simple Poisson hypothesis, should not be very different from unity. The disadvantage in such tests is that no indication is given of the nature of the departure from the null hypothesis; for these data, exploring the data structure by fitting GLMs with nested structure provides a more useful procedure for ascertaining which alternative hypothesis to the simple Poisson affords an adequate description of the data.

Chapter Seven

Random walks and diffusion processes

7.1 CONCEPTS

Interest in the properties of random walks dates from the early part of the twentieth century (Rayleigh, 1905, 1919). Lord Rayleigh (John William Strutt, third Baron Rayleigh) was a remarkable man, whose scientific curiosity was stimulated by many natural phenomena, including the shapes of pebbles, the curved flight of cricket balls, the green flash sometimes observed at tropical sunsets, and the cavitation of ships' propellers. To assist his laboratory experiments, he became a skilled glass-blower, and in later life sealed a sheet of paper and a pencil inside a glass retort, telling his family that if he wished to communicate with them from the after-life, his messages would appear on the paper inside the sealed retort. There is no record of messages having been received or detected.

The concept of a random walk will be broadly familiar from earlier chapters discussing fluctuations in water storage and dam theory. A formal description of a random walk is as follows. Imagine a particle at rest on a horizontal axis, initially (at time $t = 0$) at rest at a position X_0. At time $t = 1$, the particle makes a jump of magnitude Z_1 along the horizontal axis, where Z_1 is a random variable having a pdf $f(z)$, so that its new position on the axis is $X_1 = X_0 + Z_1$. At time $t = 2$ a further jump of magnitude Z_2 is taken, moving the particle to a new position at $X_2 = X_0 + Z_1 + Z_2$, and so on. After $t = n$ time steps, the position of the particle will be

$$X_n = X_0 + Z_1 + Z_2 + \ldots + Z_n$$

or

$$X_n = X_{n-1} + Z_n \qquad (n = 1, 2, 3, \ldots). \tag{7.1}$$

The particle therefore makes a **random walk in discrete time**. The **state space**, or set of values that X_n can possibly take, will be a discrete set if the steps are restricted to integral values (i.e. if the pdf $f(z)$ is a discrete distribution); but if $f(z)$ is a continuous distribution, then the state space will be continuous. As an example of the latter case, suppose that a reservoir had unbounded volume; then (7.1) would be the defining equation expressing storage X_n at the end of the nth wet season, in terms of storage X_{n-1} at the end of the preceding wet season, and net inflows to storage Z_n during the nth wet season.

In a **simple random walk**, the pdf $f(z)$ is discrete and takes a particularly simple form:

$$\begin{aligned} P[Z_i = 1] &= p \\ P[Z_i = 0] &= 1 - p - q \\ P[Z_i = -1] &= q \end{aligned} \tag{7.2}$$

although some texts prefer the slightly less general definition

$$P[Z_i = 1] = p, P[Z_i = -1] = 1 - p.$$

The simple random walk is the starting-point for the modelling of diffusion processes, as discussed below.

An **unrestricted random walk** occurs when the movement of the particle defined by (7.1) continues indefinitely. In many problems of practical interest, such as modelling the behaviour of reservoir contents, the random walk taken by the storage X_n is limited by one or more **barriers**, as in Chapters Four and Five. The water levels in Lake Victoria might initially considered to follow an unrestricted random walk; Figure 7.1 shows the end-of-year water levels recorded at three gauging stations on the lake (Entebbe, Jinja and Kisumu) for the period 1950–79.

The frequencies, in class intervals of 0.2 m, of the annual steps Z_i are shown in Table 7.1. In fact, it would be more sensible to work with water volumes instead of water levels, but this would require knowledge of the lake bathymetry. Furthermore, it is improbable that the annual increments Z_i could be considered as independently and identically distributed random variables, since for such a large drainage basin some part of the rain falling in one year would be stored within the basin until following years, so that the Z_i are likely to be serially correlated.

Whilst as a first approximation, water levels in Lake Victoria might be considered to come from a random walk without barriers; levels and volumes in storage of an intermittent lake, such as Lake Eyre in southern Australia, might be regarded as a random walk with a **reflecting barrier**, in the sense that when the level falls to zero it remains there until some future inflow results in positive storage. Thus Lake Eyre had plenty of water in 1890–91, but subsequently dried up; after years without any water, it began to fill in 1949 and was completely full by September 1950. The lake had almost disappeared again by 1952 (Hutchinson, 1957). Random walks with two reflecting barriers are the

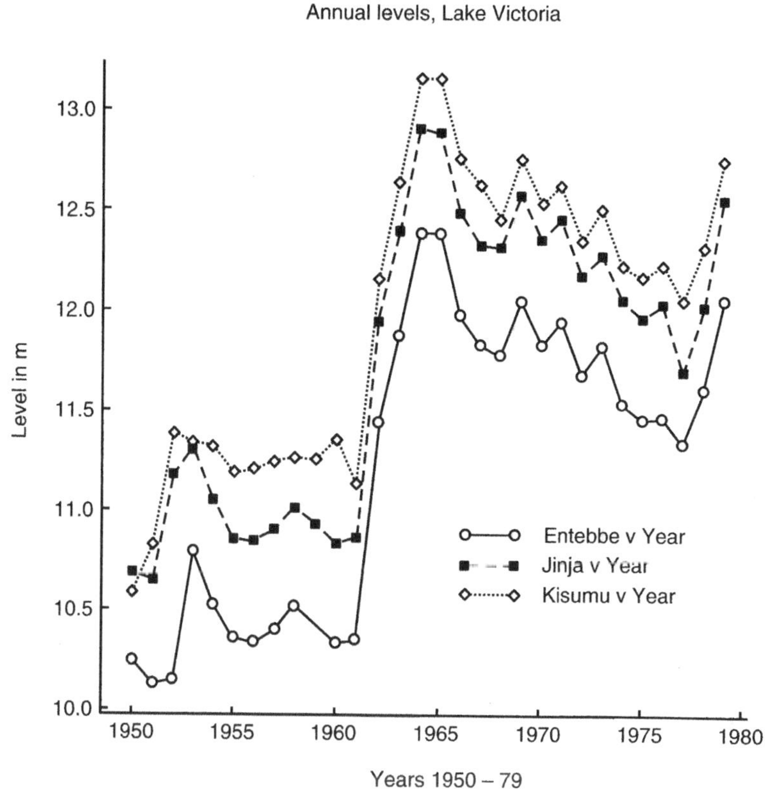

Figure 7.1 End-of-year water levels at three gauge-sites (Entebbe, Jinja, Kisumu) on Lake Victoria: 1950–79.

starting-point for models of reservoir behaviour (Chapter Three) when storage may fall to zero or attain an upper limit beyond which spillage occurs.

Where a random walk encounters reflecting barriers, interest commonly centres on the pdfs of the times taken to reach one or other of the barriers, or to reach some intermittent state, for the first time; or on the pdfs of times for which the particle remains at a barrier once it is reached; or on the pdf of the time taken to return again to a given state after entering it. At an **absorbing barrier**, on the other hand, the particle performing the random walk is retained permanently at the barrier, on reaching it; thus a sediment particle moving along a channel reach by a series of jumps, regarded as equally spaced in time, will eventually pass out of the reach, the lower end of which would then constitute an absorbing barrier.

Since the simple random walk, as defined in (7.2), is the basis for the diffusion model set out below, the following section gives some relevant results; for further development, the reader should consult Cox and Miller (1965), from whom these results are quoted.

Table 7.1 Frequencies of annual values of Z_i at Entebbe, Jinja and Kisumu

Class interval (m)	Entebbe	Jinja	Kisumu
− 0.6 to − 0.4	1	1	1
− 0.4 to − 0.2	4	6	4
− 0.2 to − 0.0	10	8	9
0.0 to 0.2	7	7	7
0.2 to 0.4	2	2	3
0.4 to 0.6	3	4	4
0.6 to 0.8	1	0	0
0.8 to 1.0	0	0	0
1.0 to 1.2	1	1	1

7.2 SOME THEORETICAL RESULTS FOR THE SIMPLE RANDOM WALK

1. For an **unrestricted random walk** defined by (7.2), and supposing the particle starts at the origin $X_0 = 0$, its position at time $t = n$ will be $X_n = Z_1 + Z_2 + Z_3 + \ldots Z_n$ which can take any of the $2n + 1$ states $0, \pm 1, \pm 2, \ldots \pm n$. If k is any one of these states,

$$P[X_n = k] = \sum n!/\{r_1!r_2!r_3!\}.p^{r_1}(1 - p - q)^{r_3}q^{r_2},$$

where the summation is over values of r_1, r_2, r_3 satisfying

$$r_1 - r_2 = k, r_3 = n - r_1 - r_2.$$

Alternatively, $P[X_n = k]$ is the coefficient of $z^k s^n$ in the expansion, in z and s, of

$$G(z,s) = z/\{-spz^2 + z[1 - s(1 - p - q)] - sq\}.$$

Since from (7.2) the mean and variance of the steps Z_i are $p - q$ and $p + p - (p - q)^2$, the mean and variance of the particle's position X_n after n steps are

$$E[X_n] = n(p - q), \text{var}[X_n] = n[p + q - (p - q)^2].$$

From the central limit theorem, the pdf of X_n is approximately

$$N(n(p - q), n[p + q - (p - q)^2]).$$

2. For a particle starting from $X_0 = 0$ and moving between **two absorbing barriers** positioned at $x = -b$ and $x = a$, eventual absorption at one or other of the barriers is certain. The probability that the particle will be absorbed by the barrier at a at time $t = n$ is the coefficient of s^n in the probability generating function

$$F_{0a}(s) = [\{\lambda_1(s)\}^b - \{\lambda_2(s)\}^b]/[\{\lambda_1(s)\}^{a+b} - \{\lambda_2(s)\}^{a+b}],$$

where

$$\lambda_1(s), \lambda_2(s) = [1 - s(1 - p - q) \pm \{\{1 - s(1 - p - q)\}^2 - 4pqs^2\}^{1/2}]/(2ps).$$

$F_{0a}(s)$ evaluated at $s = 1$ gives the probability that the particle is absorbed by the barrier at $x = a$; this is

$$F_{0a}(1) = p^a(p^b - q^b)/(p^{a+b} - q^{a+b}) \qquad p \neq q$$
$$= b/(a + b) \qquad p = q.$$

3. The probability that the particle will be absorbed by the barrier at $x = -b$ at time $t = n$, on the other hand, is the coefficient of s^n in the probability generating function

$$F_{0,-b}(s) = [\{\lambda_2(s)\}^{-a} - \{\lambda_1(s)\}^{-a}] / [\{\lambda_2(s)\}^{-a-b} - \{\lambda_1(s)\}^{-a-b}]$$

with $\lambda_1(s), \lambda_2(s)$ defined as before. The probability that the particle is absorbed by the barrier at $x = -b$ is, since absorption is certain, $1 - F_{0a}(1)$.

4. Where there is a **single absorbing barrier** at $x = a$, the probability that the particle is absorbed at time $t = n$ is the coefficient of s^n in the probability generating function

$$F_a(s) = \{\lambda_1(s)\}^{-a} = (p/q)^a\{\lambda_2(s)\}^a.$$

5. Where a particle makes a simple random walk (as defined by (7.2)) between **reflecting barriers** at $x = 0$ and $x = a$, then when $t = n$ is large the probability that the particle is in state k ($k = 0, 1, 2, \ldots a$) is

$$\pi_k = (1 - p/q)(p/q)^k/\{1 - (p/q)^{a+1}\} \qquad k = 0, 1, 2, \ldots a.$$

6. Suppose the particle makes the random walk defined by (7.2) with **one reflecting barrier** at $x = 0$. If $p > q$, the particle drifts off to infinity; suppose that $p < q$. Then when $t = n$ is large the probability that the particle is in state k ($k = 0, 1, 2, \ldots$) is

$$\pi_k = (1 - p/q)(p/q)^k \qquad k = 0, 1, 2, \ldots.$$

When the particle reaches the barrier state, it remains there for a time T with probability

$$P[T = r] = (1 - p)^{r-1}p \qquad r = 1, 2, \ldots; p < q$$

and if U denotes the duration of an excursion away from the barrier at $x = 0$, the probability that $U = n$ is the coefficient of s^n in the probability generating function $\lambda_2(s)$, defined above. When $p = q$, there is no limiting distribution; but $P[T = r] = (\frac{1}{2})^r$ for $r = 1, 2, \ldots$, and U has the probability generating function $[1 - (1 - s^2)^{1/2}]/s$, corresponding to a distribution with infinite mean.

Following this summary of selected results for the simple random walk, we now consider the case where $p + q = 1$, and the particle takes very small steps of size Δ, at very short time intervals τ. Then with

$$P[Z = +\Delta] = p, P[Z = -\Delta] = q,$$

and by the definitions of mean and variance we have

$$\mathrm{E}[Z] = \Delta(p - q), \mathrm{var}[Z] = 4\Delta^2 pq.$$

Hence, if the particle takes n very small steps, where $n\tau = t$, the position $X(t)$ of the particle after time t will have mean

$$\mathrm{E}[X(t)] = (t/\tau)\Delta(p - q)$$

and variance

$$\mathrm{var}[X(t)] = (t/\tau)4pq\Delta^2.$$

If we let Δ and τ both tend to zero in such a way that $\Delta(p - q)/\tau$ and $4pq\Delta^2/\tau$ tend to finite quantities μ and σ^2 respectively, we have a **Wiener process** or **Brownian motion process**. By the central limit theorem, the random variable $X(t)$ is normally distributed with mean μt and variance $\sigma^2 t$ and, by the properties of the random walk, $X(t) - X(0)$ and $X(u) - X(t)$ are independently distributed when $0 < t < u$.

7.3 THE DIFFUSION EQUATIONS FOR THE WIENER PROCESS

Suppose now that a particle takes small steps Δx along the x-axis in small intervals of time Δt. As before, the probability

$$P[Z = \Delta x] = p, P[Z = -\Delta x] = 1 - p.$$

Using the notation of Cox and Miller (1965), let $p(x_0, x; t)\Delta x$ be the conditional probability that the particle is at x at time t, given that it starts from x_0 at time $t = 0$. Then

$$p(x_0, x; t)\Delta x = p.p(x_0, x - \Delta x; t - \Delta t)\Delta x + q.p(x_0, x + \Delta x; t - \Delta t)\Delta x$$

and the element Δx cancels from both sides. Let $p(x_0, x; t)$ be expanded in a Taylor series in powers of Δx and Δt, and then let Δx and Δt tend to zero in such a way that $(p - q)\Delta x/\Delta t$ and $4pq(\Delta x)^2/\Delta t$ tend to finite values μ and σ^2 respectively, as before. Then the resulting equation is

$$[\tfrac{1}{2}\sigma^2]\partial^2 p(x_0, x; t)/\partial x^2 - \mu\partial p(x_0, x; t)/\partial x = \partial p(x_0, x; t)/\partial t \qquad (7.3)$$

a partial differential equation of the second order in x and the first order in t. It is the one-dimensional equation for **diffusion with an external field of force**;

$p(x_0, x; t)$ is the density of particles at the point x at time t, the coefficient $[\frac{1}{2}\sigma^2]$ is a constant associated with the medium in which diffusion occurs, and μ is a constant associated with the external force field. The equation is familiar to groundwater hydrologists in the form

$$D\partial^2 c/\partial x^2 + V\partial c/\partial x = \partial c/\partial t \tag{7.4}$$

which is the equation for the concentration $c(x, t)$ of a tracer diffusing through a porous medium with diffusion coefficient D, in a constant velocity field V.

Extension of the argument leading to (7.3) to the case of two or three dimensions is straightforward; in two dimensions, the form analogous to (7.4) is

$$D_x\partial^2 c/\partial x^2 + D_y\partial^2 c/\partial y^2 + V_x\partial c/\partial x + V_y\partial c/\partial y = \partial c/\partial t.$$

7.3.1 Random walk methods in groundwater hydrology

Equation (7.3) has been used to study the dispersion of tracers and pollutants by Monte Carlo methods. Bear and Verruijt (1987) cite a very simple particle tracking model by Prickett et al. (1981); particles are assumed to move along the x-axis with an average step length A, with a deviation from average step length lying between $\pm B$. Then the pdf of particle position after a single step is

$$\begin{aligned} p(x) &= 0 & x &< A - B \\ &= 1/2B & A - B \leqslant x &< A + B \\ &= 0 & A + B &\leqslant x \end{aligned}$$

with mean A and variance $B^2/3$.

After N such steps, assumed independent, the pdf of particle position x is approximately normal, by the central limit theorem,

$$P(x) = 1/(2\pi S^2)^{\frac{1}{2}}\exp\{-(x - M)^2/(2S^2)], \tag{7.5}$$

where $M = NA$ and $S^2 = N\sigma^2 = NB^2/3$. Equation (7.5) is compared with the solution of (7.4) in the form

$$c/c_0 = 1/(4\pi Dt)^{\frac{1}{2}}\exp\{-(x - Vt)^2/(4Dt)\}$$

giving $NA = Vt$ and $D = VB^2/(6A)$. Hence if D and V are known or can be guessed, A and B can be determined, and it is straightforward to write a computer program to calculate the positions of many simulated particles after any time t. Bear and Verruijt (1987) give an example in which particles of a pollutant are tracked within an area 400 m by 640 m, the pollutant being released 340 m upstream of a well and moving towards it at a velocity of 1 m d^{-1}. Bear et al. (1968) give an even simpler example in which particles are emitted from the origin to move along the x-axis, with independent steps of magnitude l taken in the positive or negative directions with probability $\frac{1}{2}$.

Note that the term "particle tracking" as used for example, by Konikow and Bredehoeft (1990); Bedient et al. (1990) makes no suppositions about random movement of particles following a random walk or Wiener process; instead, particles move deterministically as part of the solution of transport equations by the method of characteristics. A number of traceable particles are allocated to each cell of a finite-difference grid to form a set of points distributed in a geometrically uniform pattern throughout an area of interest. Four to nine points per cell are commonly used for two-dimensional problems. Then for each time step every point is moved a distance proportional to the length of the time increment and the velocity at the point's location. Thus the new position of a point is calculated deterministically by means of

$$x_{p,k} = x_{p,k-1} + \delta x_p = x_{p,k-1} + \Delta t V_{x[x(p,k),y(p,k)]}$$
$$y_{p,k} = y_{p,k-1} + \delta y_p = y_{p,k-1} + \Delta t V_{y[x(p,k),y(p,k)]}$$

where p is the index number of the point, and δx_p, δy_p are the distances moved in the x- and y-directions. After all points have been moved, the concentration at each node of the grid is assigned the average of the concentrations of all points then located within each grid square. The moving points simulate convective transport because the concentration at each node of the grid will change with each time step. However, the movement of particles has no probabilistic component when used in this way to solve transport equations.

7.4 DISTRIBUTION OF FIRST PASSAGE TIMES WITH ONE ABSORBING BARRIER: NUMERICAL EXAMPLE

We now give an example of how (probabilistic) particle tracking methods can be used to calculate the pdf of first passage times for particles emitted from the origin at $x = 0$, subject to a simple random walk with $P[Z = +\Delta] = p = 0.75$, $P[Z = -\Delta] = q = 0.25$ an absorbing boundary is assumed at $x = 20$, and Δ is taken to be 1. This might be considered as a model of tracer dispersion, or perhaps as model of sediment or tracer movement along a channel reach. We could obtain an analytical solution using (4) of section 7.2; however, it would be tedious to calculate the many coefficients of s^n that would be required. Instead, a simple computer program was written to track the positions of 1000 particles, taking about 20 seconds on a 486 computer. Table 7.2 gives the frequency distribution of the 1000 first passage times.

Table 7.2 Frequencies of first passage time (f_i) and numbers of time steps i

i	20	25	30	35	40	45	50	55	60
f_i	16	67	200	169	218	105	98	52	45
i	65	70	75	80	85	90	95	100	
f_i	13	8	5	1	1	2	0	0	

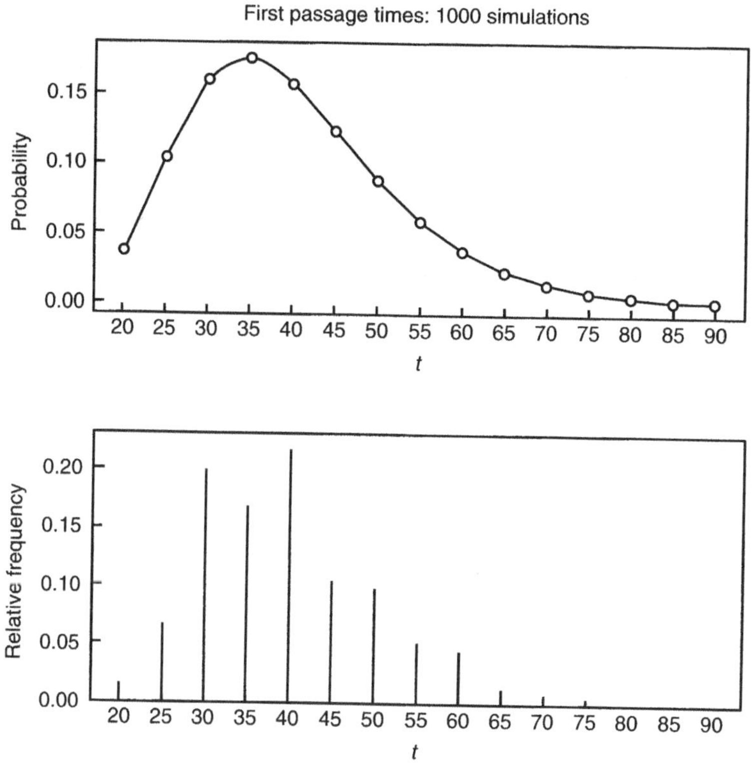

Figure 7.2 Theoretical distribution of first passage times (upper figure), and bar chart obtained from 1000 simulations (lower figure)

Suppose now we approximate the discrete random walk by a Wiener process; this we can do by taking $\sigma = 1, \mu = p - q = 0.5$. Theory shows (Cox and Miller, 1965, p. 221) that, for a Wiener process with one absorbing barrier at $x = a$, drift μ and constant σ^2, the pdf of first passage times is the inverse Gaussian distribution

$$g(t) = a/[\sigma\{2\pi t^3\}^{\frac{1}{2}}]\exp\{-(a - \mu t)^2/(2\sigma^2 t)\} \qquad t > 0 \qquad (7.6)$$

which in the present case becomes

$$g(t) = 20/\{2\pi t^3\}^{\frac{1}{2}}\exp\{-(20 - \tfrac{1}{2}t)^2/(2t)\} \qquad t > 0. \qquad (7.7)$$

Figure 7.2 shows the relative frequencies, (values in Table 7.2 divided by 1000) plotted together with the distribution in (7.7) plotted against time t; the quantities plotted are given in Table 7.3.

There is good agreement between the theoretical distribution of first passage times, and the relative frequencies obtained by particle tracking. In this example, it would have been possible to dispense with probabilistic particle tracking and use the distribution in (7.6) directly; however for more complex

Table 7.3 Ordinates of first passage time distribution (7.7), and relative frequencies $f_i/1000$

t	$f_i/1000$	Ordinate
20	0.0160	0.0366
25	0.0670	0.1036
30	0.2000	0.1601
35	0.1690	0.1762
40	0.2180	0.1577
45	0.1050	0.1233
50	0.0980	0.0879
55	0.0520	0.0587
60	0.0450	0.0373
65	0.0130	0.0229
70	0.0080	0.0137
75	0.0050	0.0080
80	0.0010	0.0046
85	0.0010	0.0026
90	0.0020	0.0015

problems where there are perhaps several absorbing boundaries in a space of more than one dimension, analytical forms for the distribution of first passage times are rarely available. Probabilistic particle tracking then provides a general method applicable to such situations.

Where particles following a Wiener process move between two absorbing barriers at $-b$ and a, Cox and Miller (1965, p. 222) show that the probability $p(x, t)$ that a particle is at position x at time t is

$$p(x,t) = \exp(\mu x/\sigma^2) \sum_{n=1}^{\infty} a_n \exp(-\lambda_n t) \sin\{n\pi(x+b)/(a+b)\}$$

where

$$\lambda_n = \tfrac{1}{2}\{\mu^2/\sigma^2 + n^2\pi^2\sigma^2/(a+b)^2\}$$

and

$$a_n = 2\sin\{n\pi b/(a+b)\}/(a+b).$$

7.5 DISTRIBUTION OF FIRST PASSAGE TIMES WITH ONE REFLECTING BARRIER: NUMERICAL EXAMPLE

Suppose now that particles emitted at the point $x = x_0 > 0$ make random walks along the positive half of the x-axis; as before

$$P[Z = +\Delta] = p, P[Z = -\Delta] = q,$$

and there is assumed to be a reflecting barrier at the origin $x = 0$. If the

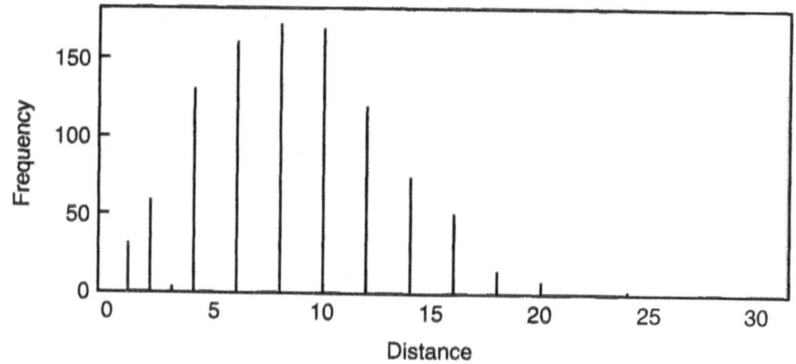

Figure 7.3 Positions of 1000 particles after $N = 25$ time-steps

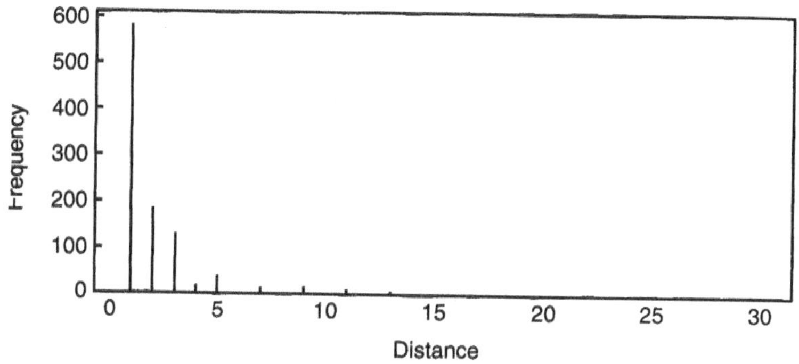

Figure 7.4 Positions of 1000 particles after $N = 50$ time-steps

probability of a positive step p is greater than or equal to the probability of a negative step q, then the probability that a particle is at x at time t tends to zero as t tends to infinity (Cox and Miller, 1965) so let us assume $p < q$; in particular, we take $p = 0.25$ and $q = 0.75$, so that the drift $\mu = p - q = -0.5$. This kind of model might be appropriate, for example, where a conservative tracer moves unidimensionally towards an impermeable barrier. A computer program to calculate the positions of 1000 points after $t = N$ time steps gave the particle positions shown in Figures 7.3 ($N = 25$) and 7.4 ($N = 50$).

For this simple case of one reflecting barrier, Cox and Miller (1965) give an analytical solution; it is

$$\begin{aligned} p(x,t) = 1/[\sigma\surd(2\pi t)].\{&\exp[-(x - x_0 - \mu t)^2/(2\sigma^2 t)] \\ &+ \exp[-(4x_0\mu t - (x + x_0 - \mu t))^2/(2\sigma^2 t)] \\ &+ (2\mu/\sigma^2)\exp[2\mu x/\sigma^2].[1 - \Phi((x + x_0 + \mu t)/(\sigma\surd t))], \end{aligned}$$

where $\Phi(x)$ is the standard normal integral. As t tends to infinity, this distribution tends to the exponential pdf

$$p(x, \infty) = 2|\mu|/\sigma^2 \exp\{-2|\mu|x/\sigma^2\} \qquad x > 0; \mu < 0$$

which is independent of x_0, the point of emission of particles. The trend towards this limiting pdf is evident from the comparison of Figures 7.3 and 7.4.

7.6 "RANDOM MEDIA" MODELS OF GROUNDWATER FLOW

Besides giving a full account of random walk models of flow through porous media, Scheidegger (1974) considers a second type of stochastic model in which deterministic flow passes through a random configuration of channels corresponding to an interconnected structure of pores. Models of this second type he calls **random media models**. They can be broadly divided into two classes: **random graph models** and **random maze models**.

Consider a small spot of tracer introduced into a porous medium; after a certain time, it will have spread through the medium as a result of the complex branching structure of the interconnected pores. The path followed by particles of the tracer will have a shape resembling a tree; Scheidegger and his colleagues (Liao and Scheidegger, 1968, 1969) took the particles of tracer to be located at the free vertices of the tree, simplified to a bifurcating process. To allow the use of probabilistic methods, a hypothetical population or **ensemble** of bifurcating trees was defined, all of which had the same given number of free ends, and all equally probable. For any one such tree, the mean distance h_{mean} of the free vertices from the tree root can be calculated, together with the variance σ_h^2 about this mean value of the distances from root to free vertices. The means, over all members of the ensemble, of h_{mean} and σ_h^2 then gave the expected value of the mean position of the tracer, and its dispersion.

Scheidegger (1974) quotes the result of Cayley (1859) giving the number N of bifurcating trees each having n free vertices:

$$N = (2n-1)^{-1}(2n-1)!/[n!(n-1)!]$$

but for large n, the factorials in this expression increase rapidly in magnitude so that it would be very difficult to enumerate all possible trees. Monte Carlo methods were used to generate a sample from the ensemble of trees.

In random maze models, the starting-point is a two- or three-dimensional lattice of interconnected nodes; call the connection between two adjacent nodes a "bond". Bonds are omitted from the maze at random; a spot of tracer introduced at a node will then spread to all nodes connected to it by an unobstructed path. However, the random omission of bonds may result in configurations in which not all nodes can be reached by the tracer so that complete saturation bcomes impossible; the random maze model was therefore modified by Torelli and Scheidegger (1971) so that, for a two-dimensional lattice, at least one of the four bonds entering each node remained. As with random graph models, an ensemble of random mazes is defined, and Monte Carlo methods are used to draw sample mazes from the ensemble.

Chapter Eight

Some computational methods

8.1 COMPUTER SIMULATION

Throughout this book, the term **simulation** has been used in different contexts. Perhaps the most common usage, familar to hydrologists, is in the context of understanding the behaviour of a complex system. For example, an interconnected set of water resources – perhaps by abstraction from several rivers and groundwater, with storage of abstracted water in reservoirs prior to distribution – the frequency and duration of failures in supply are commonly estimated by simulating the behaviour of the sources and the fluctuation in demand, taking account of the limited capacities of reservoir storages: simulation is usually necessary because the Markov chain models of reservoir behaviour given in Chapter Four are difficult to extend to many interconnected reservoirs with multiple inflows and demands. Simulation of such systems commonly requires the identification of an appropriate time-series model for the inflows, taking account of seasonal variation and cross-correlations between inflows, followed by efficient estimation of the parameters in whatever model is adopted. This is by no means straightforward, and the methods used are discussed extensively in other texts (Hipel and McLeod , 1994; Salas et al., 1980; Bras and Rodriguez-Iturbé, 1985).

A second use of simulation, mentioned in earlier chapters, is as an exploratory tool to give a feel for the kind of model which might be appropriate to describe some physical phenomenon with apparently random behaviour. As an example, consider the shot-noise model, described in Chapter Six, for the description of mean daily flows at a gauging site. If we select several sets of parameters for the shot-noise model, and calculate, for each such set, the corresponding sequence of daily flows, we will get an idea about whether the model is or is not suitable as a description of the historic sequence. By looking at the recessions in the calculated series, we will get an idea about whether some extension of the model is necessary to deal with a baseflow component; we will get an idea about which model parameters are likely to be the most critical; and we might even use the simulation procedure as a rough estimation procedure, by choosing those sets of model parameters which appeared to give the best match between the calculated series and the historic record.

Both kinds of simulation require extensive use of computer-generated pseudo-random number sequences. All statistical packages for desk-top computers now contain procedures for the simulation of sequences of pseudo-random numbers with distribution uniform over the interval (0, 1), which forms the starting-point for all other kinds of computer simulation. In addition, it would be exceedingly rare to find a reputable statistical package which, at the very least, did not contain procedures for the generation of pseudo-random numbers with the following distributions: beta, chi-squared, exponential, F, gamma, log-normal, normal, t, Weibull, binomial, hypergeometric and Poisson. Here, we give only a very few of the simplest methods, and refer the reader to specialised texts for details of the techniques by which uniformly distributed pseudo-random variables are transformed to give new pseudo-random variables with desired pdfs.

The use of simulation as a rough method of parameter estimation was mentioned in the preceding paragraph. In fact, there is a class of computer-intensive estimation procedures known as **bootstrap methods** which are often helpful when carrying out statistical tests, or for giving an idea about the variability of an estimate, in conditions where more usual statistical procedures are invalid. Section 8.3 below gives a short account of such methods; again the reader is referred to specialised texts for the details (Efron, 1982; Efron and Gong, 1983; Hall, 1992).

8.2 SUMMARY OF SOME PSEUDO-RANDOM NUMBER SIMULATION PROCEDURES

Suppose that $U_1, U_2, U_3, \ldots$ are a series of pseudo-random numbers with pdf uniform over the interval (0, 1). Then

1. $V_1, V_2, V_3 \ldots$, where $V_i = -\log_e U_i/\lambda$ are from an exponential distribution with parameter λ.
2. $V_1 + V_2 + V_3 + \ldots V_k$ have a gamma distribution with parameters k (integral) and λ: i.e.

$$f(V) = \lambda(\lambda V)^{k-1}\exp(-\lambda V)/k!.$$

3. If the cumulative sums $V_1, V_1 + V_2, V_1 + V_2 + V_3, \ldots$ are calculated, the number of such sums lying in the interval (0, 1) has a Poisson distribution with parameter λ.
4. The pseudo-random numbers

$$W_1 = (-2\log_e U_1)^{1/2}\cos(2\pi U_2)$$

and

$$W_1 = (-2\log_e U_1)^{\frac{1}{2}}\sin(2\pi U_2)$$

have pdf $N(0, 1)$; the pseudo-random numbers

$$Z_1 = \mu + \sigma W_1 \quad \text{and} \quad Z_2 = \mu + \sigma W_2$$

have pdf $N(\mu, \sigma^2)$; and the pseudo-random numbers

$$X_1 = \exp(\mu + \sigma W_1) \quad \text{and} \quad X_2 = \exp(\mu + \sigma W_2)$$

have a two-parameter log-normal distribution with parameters μ, σ.

5. If $W_1, W_2, W_3 \ldots W_k$ are pseudo-random numbers from $N(0, 1)$, then $\chi^2 = W_1^2 + W_2^2 + W_3^2 + \ldots + W_k^2$ has a χ^2 distribution with k degrees of freedom: that is, a gamma distribution with parameters $\lambda = \frac{1}{2}$ and k:

$$f(\chi^2) = \tfrac{1}{2}(\tfrac{1}{2}\chi^2)^{k-1}\exp(-\tfrac{1}{2}\chi^2)/k!.$$

6. If $W_1, W_2, W_3 \ldots W_k$ are pseudo-random numbers from $N(0, 1)$, with mean $\bar{W}$ and variance

$$s_W^2 = \sum(W_i - \bar{W})^2/(k-1),$$

then $T = \bar{W}/\surd(s_W^2/k)$ has a t-distribution with $k - 1$ degrees of freedom:

$$f(t) = \Gamma\{\tfrac{1}{2}k\}/[\surd\{\pi(k-1)\}.\Gamma(\tfrac{1}{2}(k-1))].(1 + T^2/(k-1))^{k/2}.$$

If $W_{11}, W_{12}, W_{13} \ldots W_{1k}$ and $W_{21}, W_{22}, W_{23} \ldots W_{2m}$ are two samples of $N(0, 1)$ pseudo-random numbers, sample sizes k, m, sample means $\bar{W}_1, \bar{W}_2$, sample variances

$$s_1^2 = \sum_{i=1}^{k} (W_{1i} - \bar{W}_1)^2/(k-1)$$

$$s_2^2 = \sum_{i=1}^{m} (W_{2i} - \bar{W}_2)^2/(m-1)$$

then $F = s_1^2/s_1^2$ has Fisher's F-distribution, given by

$$f(F) = C.F^{(k-3)/2}/\{((k-1)F + (m-1)\}^{(k+m-2)/2},$$

where C is a constant to make the integral of $f(F)$ equal to one, namely

$$C = (k-1)^{(k-1)/2}.(m-1)^{(m-1)/2}$$
$$\Gamma\{k-1)/2\}\Gamma\{m-1)/2\}/\Gamma\{k+m-2)/2\}.$$

8.3 SIMULATING AN INHOMOGENEOUS POISSON PROCESS IN ONE DIMENSION

Recall that for an inhomogeneous Poisson process in one dimension (say time), the rate λ varies with time, $\lambda = \lambda(t)$. The method for simulating events from an inhomogeneous Poisson process has been given by Lewis and Shedler

(1979); essentially the procedure is to generate pseudo-random numbers from a homogeneous Poisson process, and then omit some of them according to a given rule. More specifically, the procedure is as follows:

1. Calculate λ_{max}, the maximum value of $\lambda(t)$ over the period of time $(0, T)$ for which the sequence is to be generated.
2. Compute pseudo-random numbers following a homogeneous Poisson process with parameter λ_{max}, using procedure (1) of section 8.2. Thus events will have been generated which occur at times $t_1, t_2, t_3, t_4 \ldots t_N$, where $t_1 > 0$ and $t_N > T$.
3. For each such event, calculate $\lambda(t)/\lambda_{max}$, a number in $(0, 1)$. Generate a pseudo-random number U in this interval $(0, 1)$; if $U \leq \lambda(t)/\lambda_{max}$, retain the event: otherwise discard it.
4. The retained events constitute a pseudo-random sample of events from the inhomogeneous Poisson process with time varying $\lambda(t)$.

As an example, consider the generation of pseudo-random numbers from an inhomogeneous Poisson process with rate

$$\lambda(t) = 2 + \tfrac{1}{2}\cos(2\pi t/4)$$

over the interval $(0, 20)$. The rate $\lambda(t)$ is harmonic with maximum value 2.5, so we generate a sequence of pseudo-random numbers from a homogeneous Poisson process with $\lambda_{max} = 2.5$. One generated sequence of 50 values is as follows:

1.2227	0.1549	0.1547	0.0270	0.3834	0.2620	0.0741
1.3459	1.1715	0.2542	0.1595	1.9467	0.3835	1.0822
0.3494	0.1503	0.2117	0.0288	0.0668	0.2563	0.9545
0.1699	0.3508	0.1420	0.0376	0.1086	0.5351	1.2191
0.1226	0.4456	0.1831	0.1117	0.0036	0.4028	0.5593
0.0070	0.1299	0.1133	0.1714	1.0486	0.1838	0.0490
0.5197	0.3317	0.1064	0.2955	0.5746	0.5165	0.4095
0.7171						

with cumulative values:

1.2227	1.3776	1.5323	1.5593	1.9427	2.2047	2.2787
3.6246	4.7961	5.0503	5.2098	7.1565	7.5400	8.6222
8.9716	9.1219	9.3336	9.3624	9.4293	9.6855	10.6401
10.8100	11.1608	11.3028	11.3404	11.4490	11.9841	13.2032
13.3258	13.7714	13.9545	14.0662	14.0698	14.4725	15.0318
15.0389	15.1688	15.2821	15.4535	16.5021	16.6859	16.7349
17.2546	17.5863	17.6927	17.9881	18.5627	19.0793	19.4887
20.2058						

We now generate 50 uniformly distributed pseudo-random variables; for each, we test whether it is less than or equal to $(2 + \frac{1}{2}\cos(2\pi t/4))/2.5$. The result is the following 50 logical values, 1 indicating inclusion, 0 omission:

1	1	1	1	0	1	1	1	1	1	1	1
1	1	1	1	1	1	0	1	1	1	1	1
1	0	1	1	1	0	1	1	1	0	1	1
0	0	1	1	1	0	0	1	1	0	1	1
1	1										

so that the final sample is obtained by omitting from the sequence the 10 values 1.9427, 9.4293, 11.4490, 13.7714, 14.4725, 15.1688, 15.2821, 16.7349, 17.2546 and 17.9881.

8.4 SIMULATING HOMOGENEOUS AND INHOMOGENEOUS POISSON PROCESSES IN TWO DIMENSIONS

Suppose we wish to generate a realisation of events from a homogeneous Poisson process in a rectangle $0 < x < a, 0 < y < b$, with constant λ. The method of Lewis and Shedler (1979) is as follows:

1. Generate the x-coordinates of the points, by selecting points on the x-axis between 0 and a, as from a homogeneous Poisson process with parameter $b\lambda$. That is, we generate a sample $x_1, x_2, x_3, \ldots$ from the pdf $b\lambda\exp(-b\lambda x)$, calculate their cumulative sums $x_1, x_1 + x_2, x_1 + x_2 + x_3, \ldots$ stopping when the cumulative sum exceeds a.
2. Generate the y-coordinates as pseudo-random numbers from a uniform distribution over the interval $(0, b)$.

Cox and Isham (1980) give an alternative way of generating realisations of homogeneous Poisson processes in the plane. It consists of constructing a sequence of points with polar coordinates $(R_k, \Theta_k)(k = 1, 2, \ldots)$ such that

$$\pi R_1^2, \pi(R_2^2 - R_1^2), \pi(R_3^2 - R_2^2), \ldots$$

are independently and identically distributed with parameter λ, while the angular coordinates Θ_k $(k = 1, 2, \ldots)$ are generated as a sequence of independent variables each uniformly distributed on the interval $[0, 2\pi)$ and independent of the R_k.

The procedure of the last section can also be used to generate pseudo-random samples from an inhomogeneous Poisson process over a rectangle in two-dimensional space. Suppose that the rate $\lambda = \lambda(x, y)$ and that the maximum value of $\lambda(x, y)$ over the rectangle $0 < x < a, 0 < y < b$ is $\lambda_{\max}$. Then for each of the generated points, $\lambda(x, y)/\lambda_{\max}$ is calculated. Pseudo-random numbers U uniform on $(0, 1)$ are then generated, one for each of the points generated in the rectangle; if $U \leq \lambda(x, y)/\lambda_{\max}$, the point is retained, otherwise it is discarded. The remaining points are then a sample from the inhomogeneous Poisson process with rate $\lambda(x, y)$.

8.5 SIMULATING ONE-DIMENSIONAL SEQUENCES FROM A NEYMANN–SCOTT PROCESS

A clear account of the procedure for the computer generation of sequences from a Neyman–Scott process has been given by Cowpertwait (1991). We suppose that the sequence generated is of rainfall; then the procedure is as follows.

1. Times at which storms begin are generated from a Poisson process with parameter λ. That is, pseudo-random numbers $t_1, t_2, t_3 \ldots$ are generated from the pdf $\lambda\exp(-\lambda t)$; storms are then taken to begin at the times
$$t_1, t_1 + t_2, t_1 + t_2 + t_3 \ldots.$$
2. Associated with each storm is a random number N of rain-cells. To ensure that at least one rain-cell occurs in each storm, $N - 1$ is taken to be a Poisson variate with mean μ: that is, N is Poisson with mean $\mu + 1$.
3. The N_i rain-cells associated with the ith storm are initiated, following the start of the storm, after times which are exponentially distributed with parameter β. That is, N_i times $t_{i1}, t_{i2}, \ldots t_{iNi}$ are generated from the pdf $\beta\exp(-\beta t)$, and the rain-cells are then activated at times $t_{i1}, t_{i1} + t_{i2}, \ldots$ **measured from the start of the *i*th storm**.
4. The duration of the period for which a rain-cell is active, is exponentially distributed with parameter η.
5. The intensity of each rain-cell, assumed constant throughout the period for which it is active, is exponentially distributed with parameter ξ.
6. Finally, the total intensity at any instant of time t is taken to be the sum of the individual intensities of all rain-cells that are active at that time.

Cowpertwait's method, described above, is one of a number of variants of the one-dimensional Neyman–Scott process, which differ in relatively minor ways in the manner in which the depth of rain associated with each rain-cell is distributed over a time interval. In a simpler form (the Neyman–Scott white noise model, first introduced for the representation of rainfall events by Kavvas and Delleur (1975)) the intensities of rain associated with rain-cells during each storm (step (3) above) are taken to occur as instantaneous bursts occurring at the times $t_{i1}, t_{i2}, \ldots t_{iNi}$.

8.6 SIMULATING TWO-DIMENSIONAL REALISATIONS FROM A NEYMAN–SCOTT PROCESS

Model studies of the spatial growth and decay of storm cells in rainfall fields require the simulation of Neyman–Scott processes in two-dimensional space, and the method for generating two-dimensional realisations follows from the way in which the process is defined. The procedure has basically four steps: first, the generation of positions of parent events in the two-dimensional space; these parent events following a Poisson process, the method is as described in section 8.4. Second, the number of "offspring" events associated with each parent event is generated; the method described by Cressie (1993) admits the possibility that the number of offspring events can be zero, a possibility excluded in Cowpertwait's method for the one-dimensional case. Third, the offspring events are distributed around each parent event, according to a bivariate pdf $f(x, y)$, which typically might be bivariate normal. And finally, the parent events are removed, leaving the distributed offspring events.

If offspring are distributed isotropically around each parent, the required pdf will be of the form $f(r)$, where r is the radial distance from parent to offspring, and the angle θ which the vector from parent to offspring makes with the x-axis of the plane will be generated uniformly over the interval $[0, 2\pi)$. To avoid the introduction of bias from edge effects, it is necessary to include a substantial boundary region around the area for which the realisation is required; parent events and their offspring must also be generated within the boundary region.

In the most general form of the Neyman–Scott process, the Poisson process for generation of parent positions in the plane is inhomogeneous, so that the positions would be generated using the methods of section 8.4.

8.7 GENERATING REALISATIONS OF CONTINUOUS STATE VARIABLES IN THE PLANE

Chapter One drew the reader's attention to the difference in structure between data which result from a marked stochastic process in a plane (as where events originate from a Poisson process in the plane, with a "mark" or second state variable defined at each point event) and data which are observations of a single state variable varying continuously over the plane, but observed at a set of points in the plane which might lie on a regular or irregular grid or have been chosen in some more or less arbitrary fashion. Whilst a (two-dimensional) Neyman–Scott process is an example of the former, examples of the latter kind of data are: observations of depth to phreatic surface at piezometric wells; soil moisture in the top metre of soil at points on a series of transects across a small drainage basin; saturated hydraulic conductivity at points on an irregular grid. Field observations from this second kind of process, where a continuous state variable (possibly of vector form) varies continuously over two-dimensional space, are analysed using methods of geostatistics, and have been extensively described elsewhere; many key references are given in Cressie (1993) . The methodology presented there is beyond the scope of this book, but for completeness and for the purposes of the present chapter, a short account of some appropriate simulation methods is given in following sections.

8.7.1 Simulation of normal and log-normal processes in a plane: Cholesky decomposition method

Simulation is commonly restricted to processes in which the state variable Z, varying over two-dimensional space so that $Z = Z(x, y)$, is distributed normally or can be transformed to normality by some simple transformation such as the Box–Cox family

$$\begin{aligned} Z^*(x, y) &= \{Z(x, y)^\lambda - 1\}/\lambda \qquad & \lambda \neq 0 \\ &= \log_e\{Z(x, y)\} & \lambda = 0, \end{aligned}$$

where λ is a parameter to be estimated. Suppose that it is required to generate a realisation of $Z(x, y)$ at n locations $(x_i, y_i)(i = 1, 2, \ldots n)$. It is necessary to know the mean value $\mu(x, y) = E[Z(x, y)]$ and the covariance structure

$$\text{cov}[Z(x_i, y_i), Z(x_j, y_j)].$$

Corresponding to the n points in the plane, there will be a vector of means

$$\mu = [\mu(x_1, y_1), \mu(x_2, y_2), \ldots \mu(x_n, y_n)]$$

and a variance–covariance matrix $\boldsymbol{\Sigma}$ with i–jth element $\text{cov}[Z(x_i, y_i), Z(x_j, y_j)]$.

For $n \leq 1000$, the method for generating the $Z(x_i, y_i)$ uses the Cholesky decomposition of the variance–covariance matrix $\boldsymbol{\Sigma}$, by which it can be expressed as a product $\boldsymbol{\Sigma} = \mathbf{LL}^T$ where $\mathbf{L}$ is a lower triangular matrix of dimension $n \times n$. Good computer packages for statistical and mathematical calculations readily permit such a decomposition. The procedure is then as follows:

1. Calculate the matrix $\mathbf{L}$ of the Cholesky decomposition of the given matrix $\boldsymbol{\Sigma}$.
2. Calculate a vector $\boldsymbol{\varepsilon}$ of independently and identically distributed pseudo-random variables from $N(0, 1)$.
3. Calculate the $n \times 1$ vector $\mathbf{Z} = \boldsymbol{\mu} + \boldsymbol{L\varepsilon}$. Then $\mathbf{Z}$ gives the required n values $Z(x_i, y_i)(i = 1, 2, \ldots n)$.

Cressie (1993) warns that when $n > 1000$ approximately and the variance–covariance matrix $\boldsymbol{\Sigma}$ is sparse, numerical inaccuracies may result from the uncritical use of the Cholesky decomposition. Cressie gives references to methods, some apparently unpublished, with greater numerical stability for large n.

Note that the Cholesky method does not require that the stochastic process $Z(x, y)$ be stationary in the statistical sense (see e.g. Cressie, 1993). Where stationarity can be assumed, alternative methods are possible.

8.7.2 Simulation of normal and log-normal processes in a plane: the turning bands method

The turning bands method is a method for generating two- and three-dimensional realisations of Gaussian (i.e. normally distributed) state variables Z, by reducing the problem to a set of simulations in one dimension only. The method requires that the process $Z(x, y)$ be stationary and isotropic: that is,

$$\mathrm{E}[Z(x, y)] = \mu$$

with μ constant not varying with x, y; and

$$\mathrm{E}[\{Z(x_i, y_i) - \mu\}\{Z(x_j, y_j) - \mu\}] = C(h),$$

for all i and j, where h is the distance $+\sqrt{\{(x_i - x_j)^2 + (y_i - y_j)^2\}}$.

The covariance function therefore is only a function of distance h, and the covariance between $Z(x_i, y_i)$ and $Z(x_j, y_j)$ is the same as the covariance between $Z(x_i + h, y_i + k)$ and $Z(x_j + h, y_j + k)$ for all h, k. To distinguish between covariance functions in one, two and three dimensions, it is convenient to denote them by $C_1(h), C_2(h), C_3(h)$.

The turning bands method for generating $Z(x, y)$ with given covariance function $C_2(h)$ in two-dimensional space is then essentially the following:

1. Take L straight lines passing though the origin of coordinates in the plane, and let $\boldsymbol{e}_1, \boldsymbol{e}_2, \boldsymbol{e}_3, \ldots \boldsymbol{e}_L$ be unit vectors along each of the lines. The directions of the lines are generated randomly from a pdf uniform over $[0, 2\pi)$.
2. Generate one-dimensional realisations along each of the L lines, with covariance function $C_1(h)$, where the relation between $C_1(h)$ and $C_2(h)$ is defined below. Denote these generated one-dimensional realisations by $Z_1^*, Z_2^*, \ldots Z_L^*$. Each of the $Z_i^*(i = 1, \ldots L)$ will vary in value with the coordinate measured along the line.
3. To obtain the generated value $Z(x, y)$ at the point with coordinates x, y, calculate the projections of x, y on each of the L lines. (For a line making an angle θ with the x-axis, this projection is distant $x \cos\theta + y \sin\theta$ from the origin.) Take the value of $Z_1^*, Z_2^*, \ldots Z_L^*$, obtained from the L lines, at the points of projection.
4. The generated value of $Z(x, y)$ is then

$$Z(x, y) = L^{-1/2}\{Z_1^* + Z_2^* + \ldots + Z_L^*\}.$$

The difficult part of the theory of the turning bands procedure is to express $C_1(h)$, the covariance function for the one-dimensional realisations, in terms of the desired two-dimensional $C_2(h)$. A general form is (Gneiting, 1996)

$$C_1(h) = 1 + h \int_{u=0}^{h} C_2'(u)/\{h^2 - u^2\}^{-1/2} \mathrm{d}u$$

and , depending on the form of $C_2(h)$, fairly simple explicit forms of $C_1(h)$ can be obtained from this general integral. For example, if

$$C_2(h) = [1 + h^2/l^2]^{-1/2}$$

where l is a constant giving the "correlation length"), then

$$C_1(h) = [1 + h^2/l^2]^{-1}.$$

For the generation of realisations $Z(x, y, z)$ in three dimensions, there is some simplification, since (Cressie, 1993) $C_1(h)$ can be obtained from the desired $C_3(h)$ by differentiation of the product $hC_3(h)$ with respect to h.

Since realisations of this kind are marginal to the purposes of this book, we refer the reader to the growing literature on the turning bands method: a point of departure is a recent paper by Gneiting (1996).

8.8 THE BOOTSTRAP METHOD

The bootstrap is a method of providing distributional information, such as standard errors, about statistical estimates – without making precise distributional assumptions about the data. It can also provide estimates with reduced bias. This is achieved by "resampling" from the data; that is, generating new data sets by sampling with replacement from the data set being investigated. A good introduction to the bootstrap is given by Efron and Tibshirani (1986); a fuller treatment can be found in Efron and Tibshirani (1993). The method is formally as follows. Suppose we wish to obtain properties of the sampling distribution of some derived random variable $R(Z_1, Z_2, Z_3, \ldots Z_n)$ calculated from a sample of data $Z_1, Z_2, Z_3, \ldots Z_n$, **without the necessity for making assumptions about the pdf of Z_i**. Then

1. Put the $Z_i(i = 1, 2, \ldots n)$ increasing order of magnitude, giving a rearrangement denoted by $Z_{(1)}, Z_{(2)}, Z_{(3)}, \ldots Z_{(n)}$. The empirical cumulative distribution of the $Z_{(i)}$ can then be plotted, allocating an equal probability $1/n$ to each of the $Z_{(i)}$; that is, the ordered set $Z_{(i)}$ are plotted along a horizontal axis, with a step of height $1/n$ occurring at each. In symbols, if an "indicator function" $I(Z_{(i)} \le z)$ is defined such that $I(Z_{(i)} \le z) = 1$ when $Z_{(i)} \le z$ and 0 otherwise, the empirical distribution function is

$$F^*(z) = n^{-1} \sum_{i=1}^{n} I(Z_{(i)} \le z).$$

2. Draw a random sample of size n from $F^*(z)$: in effect, a sample of size n is sampled **with replacement** from the original data set $Z_i(i = 1, 2, \ldots n)$. Denote this generated sample by $Z_1^+, Z_2^+, Z_3^+, \ldots Z_n^+$; calculate the value of

$$R(Z_1^+, Z_2^+, Z_3^+, \ldots Z_n^+).$$

Denote it by R^+.

3. Repeat the calculation a large number T of times, say $T = 500$, thus obtaining T values R^+. Plot their histogram, which is an approximation to the pdf of R^+.
4. Summary statistics (such as the mean and variance of the T values) can be used to describe the properties of of R^+. In addition, a two-sided $100(1 - \alpha)\%$ confidence interval (L, U) for the mean of R^+ can be obtained by calculating values L and U, such that $100\alpha/2\%$ of the values of R^+ are less than L, and $100\alpha/2\%$ are greater than U.

References

Abrahamowitz, M. and Stegun, I.A. (1968) *Handbook of Mathematical Functions.* Dover: New York.

Back, A.J. (1997) Determinação da precipitação efetiva para irrigação suplementar através do balanço hídrico horário: um caso-estudo em Urussana S C. Ph D thesis, Universidade Federal do Rio Grande do Sul, Brazil 120pp.

Bartlett, M.S. (1960) *An Introduction to Stochastic Processes.* Cambridge University Press England.

Bear, J. and Verruijt, A. (1987) *Modelling Groundwater Flow and Pollution.* D. Reidel, Dordrecht.

Bear, J., Zaslavsky, D. and Irmay, S. (ed.) (1968) *Physical Principles of Water Percolation and Seepage.* UNESCO, Paris.

Becker, A. (1974) Applied principles of catchment simulation. In: *Mathematical Models in Hydrology* (Proc. Warsaw Symp., July (1971), vol. 2, pp. 762–74.

Bedient, P.B., Rifai, H.S., and Newell, C.J. (1990) *Ground Water Contamination: Transport and Remediation.* Prentice-Hall, New Jersey

Bernier J. (1970) Inventaire des modèles de processus stochastiques applicables à la description des débits journaliers des rivières. *Rev. Inst. Int. Statist.*, **38** (1), 49–61.

Bettes, R. and White, W.R. (1981) Mathematical simulation of sediment movement in streams. *Proc. Inst. Civ. Eng.*, **71**, 879–92.

Box, G.E.P. and Jenkins, G.M. (1970) *Time Series Analysis: Forecasting and Control.* Holden-Day, San Francisco.

Bras, R.L. and Rodriguez-Iturbé, I. (1985) *Random Functions in Hydrology.* Addison-Wesley, Reading, Mass., USA.

Bruijnzeel, L.A. (1996) Predicting the hydrological impacts of land-cover transformation in the humid tropics: the need for integrated research. In: (Eds: J.H.C. Gash, C.A. Nobre, J.M. Roberts, and R.L. Victoria) *Amazonian Deforestation and Climate*, John Wiley & Sons Ltd., Chichester.

Buishand, T.A. (1977) *Stochastic Modelling of Daily Rainfall Sequences.* Veenman and Zonen, Wageningen.

Calder, I.R. (1986) A stochastic model of rainfall interception. *J. Hydrol.*, **89**, 65–71.

Calder, I.R. (1996) Dependence of rainfall interception on drop size: 1. Development of the two-layer stochastic model. *J. Hydrol.*, **185**, 368–78.

Calder, I.R., Hall, R.L., Rosier, P.T.W., Bastable, H.G. and Prasanna, K.T. (1996) Dependence of rainfall interception on drop size: 2. Experimental determination of the wetting functions and two-layer stochastic model. Parameters for five tropical tree species. *J. Hydrol.*, **185**, 379–88.

Cayley, A. (1859) *Phil. Mag.*, **18**, 374.

Chin, E.H. (1977) Modelling daily precipitation process with Markov chain. *Water Resour. Res.*, **13**, 949–56.

Çinlar, E. (1975) *Introduction to Stochastic Processes.* Prentice-Hall, New Jersey.

Cowpertwait, P.S.P. (1991) Further developments of the Neyman–Scott clustered point process for modeling rainfall. *Water Resour. Res.*, **27** (7), 1431–38.

Cox, D.R. (1962) *Renewal Processes.* Methuen, London.

Cox, D.R. and Isham, V. (1980) *Point Processes.* Chapman and Hall, London, 188 pp.

Cox, D.R. and Miller, H.D. (1965) *The Theory of Stochastic Processes.* Chapman and Hall, London.

Cox, D.R. and Miller, H.D. (1965) *The Theory of Stochastic Processes.* Methuen, London, 398 pp.

Cox, D.R. and Lewis, P.A.W. (1966) *The Statistical Analysis of Series of Events.* Methuen, London.

Cressie, N.A.C. (1993) *Statistics for Spatial Data* (2nd. edn), Wiley, New York.

Darwin, C. (1836) *The Voyage of the Beagle.*

Eagleson, P.S. (1978a) Climate, soil and vegetation. 1. Introduction to water balance dynamics. *Water Resour. Res.*, **14** (5), 705–12.

Eagleson, P.S. (1978b) Climate, soil and vegetation. 2. The distribution of annual precipitation derived from observed storm sequences. *Water Resour. Res.*, **14** (5) 713–21.

Eagleson, P.S. (1978c) Climate, soil and vegetation. 3. A simplified model of soil moisture movement in the liquid phase. *Water Resour. Res.*, **14** (5), 722–30.

Eagleson, P.S. (1978d) Climate, soil and vegetation. 4. The expected value of annual evapotranspiration. *Water Resour. Res.*, **14** (5), 731–39.

Eagleson, P.S. (1978e) Climate, soil and vegetation. 5. A derived distribution of storm surface runoff. *Water Resour. Res.*, **14** (5), 741–8.

Eagleson, P.S. (1978f) Climate, soil and vegetation. 6. Dynamics of the annual water balance. *Water Resour. Res.*, **14** (5), 749–64.

Eagleson, P.S. (1978g) Climate, soil and vegetation. 7. A derived distribution of annual water yield. *Water Resour. Res.*, **14** (5) 765–76.

Econopouly, T.W., Davis, D.R. and Woolhiser, D.A. (1990) Parameter transferability for a daily rainfall disaggregation model. *J. Hydrol.*, **118**, 209–28.

Efron, B. (1982) *The Jacknife, the Bootstrap and other Resampling Plans.* SIAM Regional Conference Series in Applied Mathematics, 38. Philadelphia, USA.

Efron, B. and Gong, G. (1983) A leisurely look at the bootstrap, the jacknife, and cross validation. *Am. Statist.*, **37**, 3–48.

Efron, B. and Tibshirani, R. (1986) Bootstrap methods for standard errors, confidence intervals, and other measures of statistical accuracy. *Statistical Science*, **1**, 54–75.

Efron, B. and Tibshirani, R. (1993) *An Introduction to the Bootstrap.* Chapman & Hall, New York.

Einstein, H.A. (1937) Bed load transport as a probability problem. D.Sc. thesis, Federal Institute of Tech., Zurich, Switzerland.

Einstein, H.A. (1971) Probability, statistical and stochastic solutions. In: Chao-Lin Chiu (ed.), *Stochastic Hydraulics*, University of Pittsburgh, pp. 9–27.

Entekhabi D., Rodriguez-Iturbé, I. and Eagleson, P.S. (1989) Probabilistic Representation of the Temproal Rainfall process by a Modified Neyman–Scott Rectangular Pulses Model: Parameter Estimation and Validation. *Water Resour. Res.* **25**(2), 295–302.

Feller, W. (1957) *An Introduction to Probability Theory and Its Applications.* John Wiley & Sons Inc. New York.

(1957) *An Introduction to Probability Theory and its Applications* (2nd ed.). Wiley, New York.

Gabriel, K.R. and Neumann, J. (1957) On a distribution of weather cycles by length. *Quart. J. R. Met. Soc.*, **83**, 375–80.

Gabriel, K.R. and Neumann, J. (1962) A Markov chain model for daily rainfall occurrence at Tel Aviv. *Quart. J. R. Met. Soc.*, **88**, 90–5.

Gani, J. (1957) Problems in the probability theory of storage systems. *J. Roy. Stat. Soc.*, **B19**, 181–206.

Gelhar, L.W. (1974) Stochastic analysis of phreatic aquifers. *Water Resour. Res.*, **10** (3) 539–45.

Gelhar, L.W. (1984) Stochastic analysis of flow in heterogeneous porous media. In: J. Bear and M.Y. Corapcioglu (eds) *Fundamentals of Transport Phenomena in Porous Media.* M. Nijhoff, Dordrecht, Netherlands, pp. 673–718.

Gelhar, L.W. and Axness, C.L. (1983) Three-dimensional stochastic analysis of macrodispersion in aquifers. *Water Resour. Res.*, **19** (1), 161–80.

Gelhar, L.W., Gutjahr, A.L. and Naff, R.L. (1979) Stochastic analysis of macrodispersion in a stratified aquifer. *Water Resour. Res.*, **15** (6), 1387–97.

Genstat 5 Committee (1993) *Genstat 5 Release 3 Reference Manual*, Clarendon Press, Oxford.

Glasby, C.A., Cooper, G. and McGechan, M.B. (1995) Disaggregation of daily rainfall by conditional simulation from a point-process model. *J. Hydrol.*, **165**, 1–9.

GLIM 4 Reference Manual (1993) *The Statistical System for Generalized Linear Interactive Modelling.* Oxford Science Publications, England.

Gneiting, T. (1996) Comment on "A simple and efficient space domain implementation of the turning bands method" by C.R. Dietrich. *Water Resour. Res.*, **32** (11), 3391–6.

Graf, W.H. (1971) *Hydraulics of Sediment Transport.* McGraw-Hill, New York, 513 pp.

Greenwood, J.A. and Durand, D. (1960) Aids for fitting gamma distribution by maximum likelihood. *Technometrics* **2**, 55–65.

Gregory, J.M., Wigley, T.L. and Jones, P.D. (1993) Determining and interpreting the order of a two-state Markov chain: application to models of daily precipitation. *Water Resour. Res.*, **28** (5), 1443–6.

Grimmett, G.R. and Stirzaker, D.R. (1992) *Probability and Random Processes* (2nd edn). Clarendon Press, Oxford.

Hall, P.G. (1992) *The Bootstrap and Edgeworth Expansion.* Springer, New York, USA.

Hall, R.L., Calder, I.R., Nimal Gunawardena, E.R., and Rosier, P.T.W. (1996) Dependence of rainfall interception on drop size: 3. Implementation and comparative performance of the stochastic model using data from a tropical site in Sri Lanka. *J. Hydrol.*, **185**, 389–407.

Hamilton, K. and Garcia, R.R. (1986) El Niño/ Southern Oscillation events and their associated midlatitude teleconnections 1531–1841. *Bull. Am. Meteorol. Soc.* **67** (11), 1354–61.

Hershenhorn, J.S. (1984) Disaggregation of daily rainfall. M S Thesis Univ. of Arizona, Tucson Ariz., 146pp.

Hershenhorn, J.S. and Woolhiser, D.A. (1987) Disaggregation of daily rainfall. *J. Hydrol.*, **95**, 299–322.

Hipel, K.W. and McLeod, A.I. (1994) *Time Series Modelling of Water Resources and Environmental Systems.* Elsevier, Amsterdam, 1013 pp.

Hosking, J.R.M. and Clarke, R.T. (1989) Rainfall–Runoff Relations Derived from the Probability Theory of Storage. IBM Research Report: Yorktown Heights, New York 26p.

Hosking, J.R.M. and Clarke, R.T. (1990) Rainfall–runoff relations derived from the probability theory of storage. *Water Resour. Res.*, **26** (7), 1455–63.

Hosmer, D.W. and, Lemeshow S. (1989) *Applied Logistic Regression.* Wiley, Chichester, England.

Hubbell, D.W. and Sayre, W.W. (1964) Sand transport studies with radioactive tracers. *Proc. Am. Soc. Civil Eng.*, **90** HY3.

Hurley, M.A. (1992) Modelling bedload transport events using an inhomogeneous gamma process. *J. Hydrol.*, **138**, 529–41.

Hutchinson, G.E. (1957) *A Treatise on Limnology. Volume I: Geography, Physics and Chemistry*. Wiley, New York.

Ison, N.T., Feyerhem, A.M. and Bark, L.D. (1971) Wet period precipitation and the gamma distribution. *J. Appl. Meteorol.*, **10**, 658–65.

Kartvelishvili, N.A. (1969) *Theory of Stochastic Processes in Hydrology and River Runoff Regulation*. No. TT68–50478, Israel Program for Scientific Translation, Jerusalem.

Kaczmarek, Z. (1963) Fundamentals of streamflow regulation by storage reservoirs. *Arc. Hydrotech.*, **10**, 1–27.

Kavvas, M.L. and Delleur, J.W. (1975) The stochastic and chronologic structure of rainfall sequence: application to Indiana. *Tech. Rep.* 57, Water Resour. Res. Cent., West Lafayette, Indiana USA.

Khaliq, M.N. and Cunnane, C. (1996) Modelling point rainfall occurrences with the modified Bartlett–Lewis rectangular pulses model. *J. Hydrol.*, **180**, 109–38.

Klemes, V. (1978) Physically based stochastic hydrologic analysis. In: *Advances in Hydroscience*, vol. 11, pp. 285–356.

Klemes, V. (1981) Long memory flow models in reservoir analysis: what is their practical value? *Water Resour. Res.*, **17**, 737–51.

Konikow, L.F. and Bredehoeft, J.D. (1989) Computer model of two-dimensional solute transport and dispersion in ground water. *Automated Data Procesing and Computations: Techniques of Water-Resources Investigations of the US Geological Survey*, Chapter C2, Book 7.

Lewis, P.A.W. and Shedler, G.S. (1979) Simulation of non-homogeneous Poisson processes by thinning. *Naval Research Logistics Quarterly*, **26**, 403–13.

Liao, K.H. and Scheidegger, A.E. (1968) *Bull. Internat. Assoc. Sci. Hydrol.*, **13** (1), 5.

Liao, K.H. and Scheidegger, A.E. (1968) *Bull. Internat. Assoc. Sci. Hydrol.*, **14** (4), 137.

Lloyd, E.H. (1963) Reservoirs with serially correlated inflows. *Technometrics*, **5**, 85–93.

Lloyd, E.H. (1967) Stochastic reservoir theory. *Adv. Hydrosci.*, **4**, 281–339.

Lloyd, E.H. (1979) Stochastic storage problems. In: (Eds: E.H. Lloyd, T. O'Donnell and J.C. Wilkinson) *The Mathematics of Hydrology and Wate Resources*. Academic Press, London, pp. 73–85.

Lloyd, E.H. (1993) The stochastic reservoir: exact and approximate evaluations of the storage distribution. *J. Hydrology*, **151**, 65–107.

Lloyd, E.H. and Warren, D. (1992) The Gf algorithm for reservoir storage, yield and spillage distributions for non-seasonal and seasonal independent inflow processes. *Stoch. Hydraul.*, **6**, 135–49.

McLeod, A.I. and Hipel, K.W. (1995) *Reference Manual, the McLeod-Hipel Time Series Package*.

McMahon, T.A. and Mein, R.G. (1978) *Reservoir Capacity and Yield*. Elsevier Amsterdam.

McCullagh, P. and Nelder, J.A. (1989) *Generalized Linear Models* (2nd edn). Chapman and Hall, London.

Moore, R.J. (1985) The probability-distributed principle and runoff production at point and basin scales. *Hydrological Sciences Journal*, **30** (2), 273–97.

Moore, R.J. and Clarke, R.T. (1981) A distribution function approach to rainfall–runoff modelling. *Water Resour. Res.*, **17** (5), 1367–82.

Moore, R.J. and Clarke, R.T. (1982) A distribution function approach to modelling basin soil moisture deficit and streamflow. In: V.P. Singh (ed.), *Statistical Analysis*

of Rainfall and Runoff. Water Resources Publications, Colorado, pp. 173–90.
Moore, R.J. and Clarke, R.T. (1983) A distribution function approach to modelling basin sediment yield. *J. Hydrology*, **65**, 239–57.
Moran, P.A.P. (1959) *The Theory of Storage.* Methuen, London.
Negev, M. (1967) A sediment model on a digital computer. *Dep. Civ. Eng Tech. Rep. No.* 76. Stanford Univ., Stanford, California, 89 pp.
NERC Flood Studies Report (1975) Natural Environment Research Council, London.
Phartofod, R.M. (1976) Some aspects of stochastic reservoir theory. *J. Hydrol.*, **30**, 199–217.
Phartofod, R.M. (1979) The bottomless dam. *J. Hydrol.*, **40**, 337–63.
Phartofod, R.M. (1989) Riverflow and reservoir storage models. *Math. Comput. Model.*, **12**, 1057–77.
Phartofod, R.M. and Mardia, K.V. (1973) Some results for dams with Markovian inputs. *J. Appl. Prob.*, **10**, 166–80.
Popov, E.G. (1956) *Analysis of the Forming of Flow of Plain Rivers.* Gidrometeoizdat, Leningrad.
Popov, E.G. (1962) Non-uniformity of surface retention as a factor of runoff. *Bull. Int. Ass. Scient. Hydrol.*, **7**, 21–26.
Popov, E.G. (1973) Snowmelt runoff forecasts – theoretical problems. In: *The Role of Snow and Ice in Hydrology* (Proc. Banff Symp., September (1972), vol. 2, pp. 829–39.
Popov, E.G. (1980) Forecasts of seasonal inflows to reservoirs as one of the components of water control systems. In: *Hydrological Forecasting* (Proc. Oxford Symp., April 1980), pp. 279–82.
Prabhu, N.U. (1958) On the integral equation for the finite dam. *Q.J. Math.* **9**, 183–8.
Prabhu, N.U. (1964) *Time-dependent Results in Storage Theory.* Methuen London.
Prickett, J.A., Naymick, T.G. and Lonnquist, C.G. (1981) A random walk solute transport model for selected groundwater quality evaluations. *Bull. Illinois State Water Survey 65.* Champaign, Illinois.
Quinn, W.H., Zopf, D.O., Short, K.S. and Kuo Yang, R.T. (1978) Historical trends and statistics of the Southern Oscillation, El Niño and Indonesian droughts. *Fish. Bull.*, **76**, 663–78.
Raudkivi, A.J. (1967) *Losses Boundary Hydraulics.* Pergamon, Oxford, 331 pp.
Rayleigh, Lord (1905) The problem of the random walk. *Nature*, **72**, 318.
Rayleigh, Lord (1919) On the problem of random vibrations, and of random flights in one, two or three dimensions. *Phil. Mag.*, **37**, 321–47.
Rodriguez-Iturbé, I. and Valdés, J.B. (1979) The geomorphologic structure of hydrologic response. *Water Resources Res.*, **15** (6), 1409–44.
Rodriguez-Iturbé, I., Cox, D.R. and Isham, V. (1987) Some models for rainfall based on stochastic processes. *Proc. R. Soc. Lond.*, **A410**, 269–88.
Rodriguez-Iturbé, I., Cox, D.R. and Isham, V. (1988) A point process model for rainfall: further developments. *Proc. R. Soc. Lond.*, **A417**, 283–98.
Salas, J.W., Delleur, J.W., Yevjevich, V. and Lane, W.L. (1980) *Applied Modelling of Hydrological Time Series.* Water Resources Publications, Fort Collins, Colorado, USA.
Sayre, W.W. and Hubbell, D.W. (1965) *Transport and Dispersion of Labelled Bed Material, North Loop River Nebraska.* USGS Prof. Paper 433-C.
Scheidegger, A.E. (1974) *The Physics of Flow through Porous Media* (3rd edn). University of Toronto Press.
Shen, H.W. and Todrovic, P. (1971) A general stochastic model for the transport of sediment bed material. In: Chao-Lin Chiu (ed.), *Stochastic Hydraulics*, Proc. First International Symposium on Stochastic Hydraulics, Pittsburgh.
Soni, J.P. (1981) Unsteady sediment transport law and prediction of aggradation

parameters. *Water Resour. Res.*, **17** (1), 33–40.
Stern, R.D. and Coe, R. (1982) The use of rainfall models in agricultural planning, *Agricultural Meteorology*, **26**, 35–50.
Stern, R.D. and Coe, R. (1984) A model fitting analysis of daily rainfall data. *J. Roy. Statist. Soc.*, **A147**, 1–34.
Todorovic, P. and Woolhiser, D.A. (1975) A stochastic model of n-day precipitation. *J. Appl. Meteorol.*, **14**, 17–24.
Torelli, L. and Scheidegger, A.E. (1971) A review of the Stochastic Approach to Flow through Porous Media. In: *Stochastic Hydraulics* (Ed: Chao-Lin Chiu) University of Pittsburgh 654-666.
Torelli, L. and Scheidegger, A.E. (1972) *J. Hydrol.*, **15**, 23.
Tong, H. (1990) *Non-linear Time Series: A dynamical System Approach.* Clarendon Press, Oxford.
Virmani, S.M. (1975) *The Agricultural Climate of the Hyderabal Region in Relation to Crop Planning (a Sample Analysis).* ICRISAT, Hyderabad, India.
Vukmirovic, V. and Wilson, G. (Jr) (1976) Mouvement du sédiment charrié – processus aléatoire. *Proc. II International Symposium on Stochastic Hydraulics.* Lund, Sweden, 2–4 August.
Vukmirovic, V. and Wilson, G. (Jr) (1981) Mesures des caractéristiques cinématiques du sédiment charrié. *IAHR Workshop on Particle Motion and Sediment Transport.* Raperswill, Switzerland, 6–8 April.
Waymire, E. and Gupta, V.K. (1981a) The mathematical structure of rainfall representations. 1. A review of stochastic rainfall models. *Water Resour. Res.*, **17** (5), 1261–72.
Waymire, E. and Gupta, V.K. (1981b) The mathematical structure of rainfall representations. 2. A review of the theory of point processes. *Water Resour. Res.*, **17** (5), 1273–85.
Waymire, E. and Gupta, V.K. (1981c) The mathematical structure of rainfall representations. 3. Some applications of the point process theory to rainfall processes. *Water Resour. Res.*, **17** (5), 1287–94.
Weiss, G. (1973) Filtered Poisson processes as models for daily streamflow data. PhD thesis, Imp. Coll. of Sci. and Technol., London.
Weiss, G. (1977) Shot noise models for the generation of synthetic streamflow data. *Water Resour. Res.*, **13** (1), 101–108.
Woolhiser, D.A. and Pegram, G.G.S. (1979) Maximum likelihood estimation of Fourier coefficients to describe seasonal variations of parameers in stochastic daily precipitation models. *J. Appl. Meteorol.*, **18**, 34–42.
Zhao, R.J. and Zhuang, Y. (1963) Regionalisation of the rainfall runoff relations (in Chinese). *Proc. East China College of Hydraulic Engineering.*
Zhao, R.J., Zuang, Y., Fang, L.R., Lin, X.R. and Zhang, Q.S. (1980) The Xinanjiang model. In: *Hydrological Forecasting* (Proc. Oxford Symp., April (1980), pp. 351–6.

Index

Annual maximum flood, Poisson process model for 117–18
ARIMA models 5

Bartlett–Lewis models of rainfall simulation 2, 135–8
Barriers in random walk
  absorbing 63, 67, 153
  reflecting 63, 152
Bernoulli
  dependent sequences 42
  independent sequences 41
Bedload transport, Poisson model for 120–1, 138–41
Bootstrap methods 164, 172–3
Box-Jenkins models of time series 4
Brownian motion, as diffusion process 156

Chains, Markov 41
Chapman–Kolmogorov equations 80
Components of model, systematic and random error 31
Compound distributions 29
Computer simulation 9
Contributing area, distribution function of 104–5

Data 12
Determinism 2
Deviance, scaled 34–5
Deviance statistic 34, 50–1, 54
Disaggregation, of daily rainfall into hourly intervals 134–5
Distribution
  Bernoulli 23
  binomial 23
  compound 29
  conditional 22
  gamma 26
  inverse Normal 27
  negative binomial 24
  Neyman type A 25
  Normal 26
  of number of wet days in period 64
  of first passage times with absorbing, reflecting barriers 158–62
  of lengths of time in a particular state 7
  Poisson 23
  Poisson-log Normal 25
  Poisson-Pascal 25
  Polya-Aeppli 25
Dummy variable method 53

Earthquakes, and rainfall 3
Eigenvalues 45, 80
Eigenvectors 45, 80
el Niño events 14
Events
  certain 12
  disjoint 12
  elementary 12
  el Niño 14
  impossible 12
  mutually exclusive 12
Expected values, of random variables 19
Exponential family of distributions 32

Family, exponential, of distributions 32
Field 12
First passage time, mean 78
First passage times 7

Flood, annual maximum 116–17
Flood peaks, Poisson model for 116–17
Frequency relationships, between runoff and rainfall 105–7
Function
  likelihood 34
  link in GLMs 32
  variance 33

Gamma distribution, shape parameter 59
Generalised Linear Model 30, 34
Generating functions, probability 27
Geomorphological structure of hydrological response 126–30
Geostatistics 8
GLM (Generalised Linear Model) 30
Goodness of model fit 34
Groundwater hydrology, diffusion processes in 157–8

Independent Bernoulli trials 23
Independent events 17
Interception, of rainfall by leaf-surfaces 119–20

Jerimum reservoir, NE Brazil 88–90

Lake Eyre 152
Lake Victoria 152–3
Likelihood, maximum 34
Link function, in Generalised Linear Models 32
Logistic regression 48, 50, 66

Markov chains 7, 9, 41
  for modelling soil water storage 77
  second-order 46, 51
Markov process 42
Mean first passage time 78
Mean recurrence time 78
Model components, systematic and random error 31
Model
  current 35
  full 35
  maximal 35
  minimal 35
  null 35
  uses 55
Models:
  ARIMA 5
  Bartlett–Lewis, rainfall simulation 2
  Generalised Linear 30
  Markov chain 7, 9
  mathematical 12
  Neyman–Scott, rainfall simulation 2
  Stern–Coe, daily rainfall simulation 2, 9
  stochastic 5
  time series 4
Monte Carlo tests 146–9
Moran model of storage 84–8

Neyman–Scott process, simulation of 168–9

Parameter, scale, in GLMs 33
pdf (probability density function)
  Bernoulli 23
  binomial 23
  compound 29
  gamma 26
  negative binomial 24
  Neyman type A 25
  Normal 26
  Poisson 23
  Poisson-logNormal 25
  Poisson-Pascal 25
  Polya-Aeppli 25
Poisson model, testing validity of 141–6
Poisson process 24, 111–49
  combination of several processes 114
  for flood peaks over a threshold 115–17
  for modelling annual rainfall 123–5
  hydrological response and geomorphology 126–30
  interception of rainfall 119–20
  modelling bedload transport 120–1
  modelling hourly rainfall 130–4
  multiple occurrences 115
  shot-noise models 121–3
  with time-dependent rate constant 113
Probabilities, of transitions between states 42–3, 46
Probability, definition 14
  Central Limit Theorem 27
  conditional 15
  conditional distributions 22
  density function 18
  distribution function 18

of dry spells in a period 66
generating functions 27
Independent events 15, 17
Theorem of Total 16
Probability distribution 7
Probability theory, role of 12–13
Processes
alternating renewal 126
Bartlett–Lewis 26
marked point 117
Markov, first-order 42
Neyman–Scott 27
Poisson 9, 24, 112–49
random walk and diffusion process 151–62
renewal 125–6
spatial temporal 9
stochastic 6
Pseudo–random number generation 164–5

Rainfall, modelling occurrence of 46
Rainfall depth, modelling of 56
Rainfall-runoff modelling, stochastic reservoir theory in 95–109
Rainfall-runoff relationships 9
Rainfall, relation with earthquakes 3
daily, at Manaus 49
Random graph models 162
Random maze models 162
Random media models, of porous media 162
Random variables 5, 18
expected values of 19
Random walk, and diffusion processes 151–61
unrestricted 152
Realizations 7
Renewal processes 125–6
alternating 126
Recurrence relations 55, 62
Recurrence time, mean 78
Regression, logistic 48, 50, 66
Reservoir behaviour 9
Reservoir sotage, integral equation for 90–2
Routing, of runoff 99–100
Runoff, generated, distribution function of 103–4
Runoff routing 99–100

Sample spaces 12
Scale parameter, in GLMs 33
Scaled deviance 34–5
Seasonality, modelling of 46
Sediment yield, modelling of 108
Shot-noise model, for modelling streamflow sequences 121–3
Simulation, computer 9
of artificial rainfall sequences 73
of normal and lognormal processes in a plane 170
turning bands method 171–2
Soil water content, pdf of 69
Spatial process, Poisson model in 113
Spatial-temporal process 9
State space 6
State variables 8
States, in Markov process 42
Stern-Coe model of daily rainfall 2, 9
Stochastic processes 6
marked 8
Storage models 77–93
Survivor function 128

Transition probabilities 42, 46
n-step 45
time invariant 44
Time series models 4
Time trends, incorporation into Stern–Coe models 72–3
Turning bands method, for simulating planar processes 171–2

Uncertainty and determinism 2

Variable, dummy, use for comparing regressions 53
Variance function, in GLMs 33

Wiener process 156

Printed and bound by CPI Group (UK) Ltd, Croydon, CR0 4YY

16/09/2026

14963471-0001